ENVIRONMENTAL SCIENCE, ENGINEERING AND TECHNOLOGY

VOLATILE ORGANIC COMPOUNDS

ENVIRONMENTAL SCIENCE, ENGINEERING AND TECHNOLOGY

Additional books in this series can be found on Nova's website under the Series tab.

Additional E-books in this series can be found on Nova's website under the E-books tab.

ENVIRONMENTAL SCIENCE, ENGINEERING AND TECHNOLOGY

VOLATILE ORGANIC COMPOUNDS

JONATHAN C. HANKS
AND
SARA O. LOUGLIN
EDITORS

Nova Science Publishers, Inc.
New York

For permission to use material from this book please contact us:
Telephone 631-231-7269; Fax 631-231-8175
Web Site: http://www.novapublishers.com

Additional color graphics may be available in the e-book version of this book.

Library of Congress Cataloging-in-Publication Data

Volatile organic compounds / editors, Jonathan C. Hanks and Sara O. Louglin.
p. cm.
ISBN 978-1-61324-156-1 (hardcover)
1. Volatile organic compounds--Environmental aspects. I. Hanks, Jonathan C. II. Louglin, Sara O.
TD885.5.O74V655 2011
628.5'2--dc22

2011010111

Published by Nova Science Publishers, Inc. † New York

Contents

PREFACE

Volatile organic compounds (VOCs) refers to organic chemical compounds which have significant vapor pressures. This new book presents current research in volatile organic compounds, including the effect of transition metals on the reductive dechlorination of COCs by iron-bearing soil minerals; photocatalytic degradtion of VOC gases using a short wavelength UV light and water droplets; catalytic incineration of VOCs; transport of VOCs in polymers; sources and elimination of VOCs and in-vivo analysis of palm wine VOCs by proton transfer reaction-mass spectrometry. (Imprint: Nova press)

Chapter 1 - Chlorinated organic compounds (COCs), known as volatile organic compounds (VOCs), are widespread groundwater and soil. Due to their persistence in natural environments and toxicity to humans and animals, the efforts to treat the contaminants have been continuously made to date. In this chapter, iron-bearing soil minerals (pyrite, magnetite, green rust, iron sulfide (FeS)) and iron-bearing phyllosilicates were examined to reductively dechlorinate the COCs (tetrachloroethylene (PCE), trichloroethylene (TCE), *cis*-dichloroethylene (*cis*-DCE), vinyl chloride (VC), 1,1,1-trichloroethane (1,1,1-TCA), carbon tetrachloride (CT)). Dechlorination kinetics were described by a modified Langmuir-Hinshelwood model. The rate constant for reductive dechlorination of chlorinated ethylenes (PCE, TCE, *cis*-DCE, and VC) by pyrite (0.98 – 1.71 d^{-1}) was greatest, followed green rust (0.59 – 1.59 d^{-1}), magnetite (0.19 – 0.25 d^{-1}) and phyllosilicates (0.08 – 0.40 d^{-1}). The surface area-normalized pseudo-first-order rate constants for chlorinated ethylenes by green rust were 3.4 – 8.2 and 62 – 290 times greater than those by pyrite and magnetite, respectively. The calculated specific reductive capacity of green rust for chlorinated ethylenes was in the range of 9.86 – 18.0 μM/g, which was greatest among iron-bearing soil minerals (pyrite: 1.01 – 2.26 μM/g;

magnetite: 0.33 – 0.73 μM/g; phyllosilicates: 0.18 – 1.06 μM/g). Acetylene and small amount of chlorinated intermediates were produced during the reductive dechlorination of chlorinated ethylenes by pyrite and magnetite while no chlorinated intermediates were observed throughout the dechlorination of chlorinated ethylenes by iron-bearing phyllosilicates. The effect of transition metals (Cu(II), Co(II), Ni(II), and Pt(IV)) on the reductive dechlorination of COCs by iron-bearing soil minerals was also characterized in this chapter. The rate constants for the dechlorination of 1,1,1-TCA were significantly increased by addition of 10 mM of Co(II) and Ni(II) in FeS suspension compared to those by FeS alone. In addition, the dechlorination of PCE was greatly enhanced by addition of Pt(VI) in green rust suspension, resulting in 2 orders of magnitude higher rate constant rather than that of green rust alone. X-ray photoelectron spectroscopy analysis showed that the production of zero-valent forms (Ni(0) and Pt(0)) played a catalyst role to accelerate the reductive dechlorination of COCs.

Chapter 2 - Volatile organic compounds (VOCs) are of environmental concern because of their adverse effects on human health. Moreover, VOCs can adversely affect certain manufacturing processes; for example, in semiconductor manufacturing, the wafer surface is damaged by VOCs. Although chemical filtration of classical removal techniques are effective in removing a variety of hazardous VOCs from air, these methods are expensive, and the materials used for adsorption and filtration have short and unpredictable life spans. Accordingly, there is much interest in developing new air purification systems for removing VOCs from air. In this Chapter, introduces the photodegradation of toluene and benzene with TiO_2 and $UV_{254+185nm}$ irradiation from an ozone-producing UV lamp as a first step to show VOC degradation in gas phase using short-wavelength UV irradiation with TiO_2 catalyst with different UV sources. The results show that VOCs were decomposed and mineralized efficiently owing to the synergetic effect of photochemical oxidation in gas phase and photocatalytic oxidation on the TiO_2 surface. The conversion levels obtained with $UV_{254+185nm}$ photoirradiated TiO_2 were much higher than those obtained with conventional UV sources (UV_{365nm} and UV_{254nm}), which suffer from both catalyst deactivation and the generation of harmful intermediates. The products from the photodegradation of VOCs with the $UV_{254+185nm}$ photoirradiated TiO_2 were mainly mineralized CO_2 and CO, but some water-soluble organic intermediates were also formed under more severe reaction conditions. The water-soluble aldehydes and carboxylic intermediates disappeared from the effluent gas stream and were detected in the water impingers. These findings suggest that the intermediates can be washed out by conventional gas washing technique, such as wet scrubber or air washer. As a second step, removal of water-insoluble gaseous pollutants (NOx and toluene) is introduced with $UV_{254+185nm}$ irradiation in humid air to provide a useful process for effectively removing gaseous pollutants from the air and the treatment of intermediates by trapping the water-soluble intermediates into water droplets by the air washer. The results show that the OH radicals and ozone produced by $UV_{254+185nm}$ irradiation effectively degraded NOx and toluene to HNO_3 and CO_2, respectively. The organic intermediates formed during toluene degradation were highly water soluble and could therefore be effectively removed along with the HNO_3, by the air washer. However, using the air washer as a means for effective removal of gaseous pollutants and their intermediates has disadvantages for example, 1) the reaction can be completed with 2-step process and 2) the size of air washer is relatively large. For these reasons, VOCs degradation using an ultrasonic mist generated from TiO_2 suspension under UV_{365nm}, UV_{254nm}, and $UV_{254+185nm}$ irradiations is studied and described. With this technology, gaseous pollutants and the intermediates could be degraded and captured by water droplet and decomposed further in a liquid phase with 1-step process by generating mist droplets from ultrasonic atomization of TiO_2 suspension. Organic gas species, UV wavelength, the diameter of ultrasonic mist containing photocatalyst particles (UMP), and the inhibition of the formation of secondary particles from intermediates were also investigated. In this method, VOCs were decomposed on the surface or inside of UMP and the degradation intermediates from hydrophobic gas could also be captured by water droplet and decomposed further in a liquid phase.

Chapter 3 - Volatile organic compounds (VOCs) make up a major class of air pollutants. This class includes pure hydrocarbons, partially oxidized hydrocarbons (organic acids, aldehydes, ketone), as well as organics containing chlorine, sulfur, nitrogen, or other elements. These compounds are usually found in most manufacturing processes, either for the raw materials, intermediates, or the finished products. Organic materials are present as chemicals, solvents, release agents, coatings, decomposition products, pigments, and so on that eventually must be disposed. Catalytic incineration is a well known process to destruct VOC emissions in air at low energy cost; it is useful when these contaminants are toxic and malodorous. The advantages of catalytic oxidation can be important mainly because of potential savings following the lower temperatures required. The environmental impact can be improved because both higher efficiency of abatement and lower levels of NO_x and CO_2 emissions can be reached. Furthermore the small pressure drops make the process very attractive. This article aims to show and discuss the results from continuous tests conducted in the laboratory scale reactors. Measurements of the abatement in the simulated industrial conditions, the assessment of the catalyst aging, and the identification of the possible poisoning will be shown. The catalyst powders were prepared by the incipient wetness impregnation method with aqueous solutions of metal nitrate and calcined at proper temperatures. The finished catalysts were characterized first by DTA–TGA. The effects of operating parameters, such as inlet temperature, space velocity, VOCs inlet concentration, and oxygen concentration on the catalytic incineration of VOCs over the catalysts were then performed. The activity of the catalyst decreased significantly with time while VOCs incineration was operated under a low temperature. However, the activity of the catalyst did not change much while the operating temperature was high. The catalysts were characterized by the surface and pore size analysis, XRD, XPS, EDS and SEM before and after the tests. Three kinetic models (i.e., the power–rate law, the Mars and Van Krevelen model, and the Langmuir–Hinshelwood model) were used to analyze the results. A differential reactor design was used for best fit of kinetic models in this study.

Chapter 4 - Leaking of volatile organic compounds (VOCs) from gasoline during its storage, handling and transportation constitutes a serious ecological problem seeing that VOCs are known as toxic, environmentally harmful and carcinogenic agents. Despite the fact that the lost amounts of mainly hydrocarbons during common operations in refineries or at fuel stations may seem negligible; in reality they reach hundreds of tons per year of valuable industrial products. All above mentioned facts are the reason why the separation of these compounds from air and their recycling is critically important. At the present time, VOCs removal from the air is realized by traditional cost-consuming technologies like adsorption or refrigeration and by ecological progressive high-efficiency membrane separations. Polymer membranes based on polydimethylsiloxane (PDMS) or polyether-block-amide (PEBA) currently belong to the group of polymers used for the preparation of composite separation membranes [1-4]. Some of their unfavourable limitations (lower chemical resistance, swelling) led researchers to test also other potentially utilisable polymers like polyvinylidene fluoride (PVDF), poly{1-trimythylsilyl-1-propyne} (PTMSP), high free volume amorphous glassy perfluoropolymers (Teflon AF) or cross-linked poly(amide-imide) polymers[2-4]. Hence, detailed knowledge of polymer structure-permeability relationship and polymer-penetrant interactions plays an important role in the development and potential industrial application of newly prepared membrane materials.

Generally, the mass transport of VOCs in and through the polymer matrix is a complex process which depends on polymer properties (glassy/rubbery state, orientation, porous/nonporous structure, symmetric/asymmetric architecture etc.), penetrant properties (molecular size and shape, specific penetrant-penetrant or penetrant-polymer interactions) and also on external conditions (temperature, pressure, concentration gradient etc.). It is generally accepted that mass transport in dense polymer membranes takes place according to the well known solution-diffusion mechanism (SDM) [5]. For small, not-self-aggregative, low-sorbed molecules SDM is valid without any limitations. In other cases, especially for VOCs the solubility of the compound has a strong influence on the polymer behaviour (swelling, plasticization, chain flexibility, reorganisation of dynamic free volume elements) and, consequently, on diffusivity and permeability [1, 5]. Therefore the concentration-dependence of transport parameters must be taken into account. Chapter 4 gives a survey of the VOCs transport in non-porous polymer membranes with special reference to the phenomenon of concentration-dependence of transport.

Chapter 5 - The volatile organic compounds (VOCs) are defined, according to USEPA, as those organic compounds that at 20 °C present a vapour pressure equal or higher than 0.01 KPa. This defines about 200 chemical compounds such alcohols, chlorinated hydrocarbons, esters, etc. excluding CH_4. The European Union defines VOCs as any organic compounds that has a starting boiling point lower or equal to 250 °C measured in standard atmospheric pressure (101.3 kPa). VOCs are harmful due to different factors; on the hand they form part of the photochemical smog and some of them participate in the greenhouse effect and the other, some of them are cancerous or teratogenics. VOCs sources are classified in biogenic, created by nature, and anthropogenic mainly linked to transport and use of solvents. The control of VOCs can be divided in two groups: primary, which is related to technological substitution and secondary, which are related to the elimination at the end of pipe. The last technologies are also classified in recovery methods (adsorption, condensation, membranes) and destruction methods (biofiltration, thermal and catalytic oxidation).

Chapter 6 - The in-vivo volatile organic compounds (VOCs) release patterns in palm wine was carried out using the PTR-MS. In order to analyze the complex mixtures of VOCs in palm wine, the fragmentation patterns of 14 known aroma compounds of palm wine were also investigated. Results revealed masses m/z (43, 47, 61,65,75,89 and 93) as the predominant ones measured in-breathe exhaled from the nose, during consumption of palm wine. Further studies of aroma's fragmentation patterns, showed that the m/z 43 is characteristic of fragment of various compounds, while m/z 47 is ethanol, m/z 61(acetic acid), m/z 65 (protonated ethanol cluster ions), m/z 75 (methyl acetate), m/z 89 (acetoin) and m/z 93 (2-phenylethanol) respectively. The dynamic release parameters (I_{max} and t_{max}) of the 7 masses revealed significant ($P=0.05$) differences, between maximum intensity (I_{max}) and no significant ($P=0.05$) differences between t_{max} among VOCs respectively.

In: Volatile Organic Compounds
Editors: J. C. Hanks et al. pp. 1-45
ISBN 978-1-61324-156-1

Chapter 1

REDUCTIVE DEGRADATION OF VOLATILE ORGANIC COMPOUNDS BY IRON-BEARING SOIL MINERALS AND PHYLLOSILICATES

Sungjun Bae, Youngho Sihn and Woojin Lee*
Dept. of Civil and Environmental Engineering, Korea
Advanced Institute of Science and Technology, 373-1 Daejeon, Korea

ABSTRACT

Chlorinated organic compounds (COCs), known as volatile organic compounds (VOCs), are widespread groundwater and soil. Due to their persistence in natural environments and toxicity to humans and animals, the efforts to treat the contaminants have been continuously made to date. In this chapter, iron-bearing soil minerals (pyrite, magnetite, green rust, iron sulfide (FeS)) and iron-bearing phyllosilicates were examined to reductively dechlorinate the COCs (tetrachloroethylene (PCE), trichloroethylene (TCE), *cis*-dichloroethylene (*cis*-DCE), vinyl chloride (VC), 1,1,1-trichloroethane (1,1,1-TCA), carbon tetrachloride (CT)). Dechlorination kinetics were described by a modified Langmuir-Hinshelwood model. The rate constant for reductive dechlorination of chlorinated ethylenes (PCE, TCE, *cis*-DCE, and VC) by pyrite (0.98 – 1.71 d^{-1}) was greatest, followed green rust (0.59 – 1.59 d^{-1}), magnetite (0.19 – 0.25 d^{-1}) and phyllosilicates (0.08 – 0.40 d^{-1}). The surface area-

* Corresponding author : phone : +82-42-350-3624 ; fax : +82-42-350-3610; email: woojin_lee @kaist.ac.kr

normalized pseudo-first-order rate constants for chlorinated ethylenes by green rust were 3.4 – 8.2 and 62 – 290 times greater than those by pyrite and magnetite, respectively. The calculated specific reductive capacity of green rust for chlorinated ethylenes was in the range of 9.86 – 18.0 μM/g, which was greatest among iron-bearing soil minerals (pyrite: 1.01 – 2.26 μM/g; magnetite: 0.33 – 0.73 μM/g; phyllosilicates: 0.18 – 1.06 μM/g). Acetylene and small amount of chlorinated intermediates were produced during the reductive dechlorination of chlorinated ethylenes by pyrite and magnetite while no chlorinated intermediates were observed throughout the dechlorination of chlorinated ethylenes by iron-bearing phyllosilicates.

The effect of transition metals (Cu(II), Co(II), Ni(II), and Pt(IV)) on the reductive dechlorination of COCs by iron-bearing soil minerals was also characterized in this chapter. The rate constants for the dechlorination of 1,1,1-TCA were significantly increased by addition of 10 mM of Co(II) and Ni(II) in FeS suspension compared to those by FeS alone. In addition, the dechlorination of PCE was greatly enhanced by addition of Pt(VI) in green rust suspension, resulting in 2 orders of magnitude higher rate constant rather than that of green rust alone. X-ray photoelectron spectroscopy analysis showed that the production of zero-valent forms (Ni(0) and Pt(0)) played a catalyst role to accelerate the reductive dechlorination of COCs.

I. INTRODUCTION

Volatile organic compounds (VOCs) refers to organic chemical compounds which have significant vapor pressures. Chlorinated organic compounds (COCs) such as chlorinated ethanes and ethylenes are one of the VOCs groups which can significantly affect to the environment and human health. Four types of chlorinated ethylenes in particular can be classified depend on the number of chloride in ethylene structure (tetrachloroethylene (PCE), trichloroethylene (TCE), 1,1-dichloroethene, 1,2-(*cis and trans*)-dichloroethylenes (DCEs), and vinyl chloride (VC)). PCE and TCE have been used for metal degreasing, solvent clear and drycleaning fluids. In addition, DCEs and VC have been used as a solvent for waxes, resins, polymers, fats, and production of polymer polyvinyl chloride, respectively. Their persistence in natural environments and toxicity to animals and humans have been well-known in everywhere [1-3]. For example, Substantial research on the biodegradation of chlorinated organics has been conducted for the last few decades due to their resistance to microorganisms lived in natural environment [1,4]. Therefore, the efforts to treat and degrade chlorinated ethylenes have been extensively studied using biotic and abiotic remediation technologies [4-8]. Early research was mainly focused on the biotic reductive degradation of chlorinated organics, which generally includes dechlorination mechanism such as halorespiration [5,6,8-12] and cometabolism [7,13]. However, high contaminant concentrations and low temperatures limit the reactivity of microorganisms [14]. Also, sequential transformation products (e.g., TCE, DCEs and VC) accumulated by the microbial reductive dechlorination have been reported to be more toxic and persistent than PCE [15].

Due to the production of more toxic daughter compounds and slow degradation rate during biotic remediation, abiotic reductive dechlorination has attracted an attention. Natural attenuation and in-situ redox manipulation [12,16] are examples of attractive remediation techniques for the abiotic reductive dechlorination processes because they can transform contaminants by natural reductants contained reactive sources such as Fe(II) and sulfur. Especially, the reactivity of iron-bearing soil minerals abundant in natural environments has been investigated as a natural reductants. Abiotic transformation of carbon tetrachloride (CT), trichloroethane (TCA), TCE, and PCE by iron sulfides has been investigated using pyrite [17,18], troilite [19], and mackinawite [20,21]. The reactivity of sulfide minerals could be caused by Fe(II) or sulfide on surfaces of the minerals, depending on the experimental conditions employed [18,19]. Also, CT, hexachloroethane (HCA), penta-chloroethane, and PCE have been shown to undergo reductive transformation in the presence of green rust (GR), which is a layered Fe(II)-Fe(III) hydroxide solid with a different type of anions such as Cl^-, SO_4^{2-}, CO_3^{2-}, and F^- in its interlayer [22-24].

It has been reported that iron-bearing phyllosilicates (biotite and vermiculite) in the presence of sulfide showed the 10% of CT degradation. This reaction is caused by electron transfer from Fe(II) in phyllosilicates, from sulfide adsorbed onto the phyllosilicates, and from secondary iron sulfides produced by sulfide and dissolved Fe(II) [25].

Recently, considerable effort has been made to enhance the reactivity of reductants used for the reductive degradation of chlorinated compounds. Transition metals such as Pd, Pt, Ni, and Cu were added to zero-valent metals for this purpose [26-28]. The enhanced degradation rate of chlorinated aliphatic and aromatic compounds by zero-valent iron (ZVI) and zero-valent zinc has been known by addition of Pd [29,30]. Ni and Cu have been coated on the surfaces of ZVI to increase the kinetic rate of the reductive degradation of 1,1,1-trichloroethane (1,1,1-TCA) [26].

The modification of GRs by the reactive additives has been also explored for the enhancement of reductive dechlorination by GRs. A substantial improvement in the reductive dechlorination by GRs was observed when Ag(I), Au(III), and Cu(II) were added to GR-SO_4 and GR-Cl [31-33]. The transition metals were bound on GR surfaces and reduced to a zero-valent state. They were believed to act as a catalyst by facilitating the electron transfer from GRs to chlorinated organic compounds [31].

The characterization of reductive dechlorination of COCs by iron-bearing soil minerals is important because it can produce basic knowledge that could be used to predict the fate of COCs in natural environments and to effectively operate remedial technologies.

The effect of transition metal on the reactivity of iron-bearing soil minerals to COCs is also needed to accelerate the reaction kinetic and minimize the generation of toxic products during the contaminants degradation. However, no significant efforts have made to characterize the reductive dechlorination of COCs by iron-bearing soil minerals. And, few significant studies have been carried out to fully investigate the effect of transition metals on the reductive dechlorination of COCs by iron-bearing soil minerals. Research has been carried out in this chapter 1) to investigate the reductive capacity of iron-bearing soil minerals (magnetite, pyrite, and GR) and phyllosilicates (biotite, vermiculite, and montmorillonite) for COCs (chlorinated ethylenes), 2) to characterize the degradation of chlorinated ethylenes by iron bearing soil minerals and phyllosilicates, 3) to identify the reaction mechanisms of two objectives above, 4) to investigate the effect of transition metals (Cu(II), Co(II), Ni(II), and Pt(IV)) on the reductive dechlorination of COCs (PCE, CT, and 1,1,1-TCA) by iron-bearing soil minerals (GRs and FeS), and 5) to identify the reaction mechanisms of enhanced reductive dechlorination of COCs by addition of transition metals.

II. Materials and Methods

The all chemical reagents and samples were prepared and kept in the anaerobic chamber (Coy Laboratory Products Inc., 95% N_2 and 5% H_2). Deaerated deionized water (DDW) was prepared by purging 18 MΩ•cm deionized water with 99.99% N_2 gas for 2 hrs. DDW was used for the preparation of aqueous solutions, the synthesis of soil minerals, and chemical reagents.

II.1. Chemicals

Target organic stock solutions were prepared by diluting them in methanol. Chemicals used in the experiment include: carbon tetrachloride (CT, 99.5%, Junsei), chloroform (CF, 99%, Junsei), methylene chloride (MC, 98.0%, Junsei), 1,1,1-trichloroethane (1,1,1-TCA, 99 + %, Sigma–Aldrich), 1,1-dichloroethane (1,1-DCA, 200 μg/mL, Supelco), PCE (99.9%, Sigma), trichloroethylene (TCE, 99.6%, Sigma), cis-dichloroethylene (c-DCE, 97.0%, Sigma), trans-dichloroethylene (trans-DCE, 98%, Sigma), 1,1-dichloroethylene (1,1-DCE, 99.0%, Sigma), vinyl chloride (VC, 20,000 mg/L, Sigma), methanol (99.8%, HPLC grade, EM), ferrous chloride (tetrahydrate, 99%, Aldrich), ferrous sulfate (heptahydrate, 99+%, Aldrich), $FeSO_4 \cdot 7H_2O$ (102.8%, Sigma), Na_2S (98+%, Sigma–Aldrich), $Na_2S_2O_4$ (88%, Sigma), $NaHCO_3$ (100.3%, Sigma), H_2SO_4 (95.7%, Sigma), NaOH (97.0%, EM), sodium fluoride (99.8%, J.T. Baker), platinum chloride ($PtCl_4$, 98%, Aldrich), $CuCl_2$ (97%, Junsei), $NiCl_2$ (96%, Junsei), and $CoCl_2$ (97%, Junsei). Standard gases of ethane (99.0%), ethylene (1.0%), and acetylene (1000 mg/L) (Scott Specialty Gases) were used for the analysis of non-chlorinated transformation products. A $NaHCO_3$ (10mM) buffer, acid, and base solutions were prepared by adding an exact amount of $NaHCO_3$, H_2SO_4, and NaOH to DDW, respectively.

Biotite (Bancroft, Canada), vermiculite (Transvaal, South Africa), montmorillonite (Gonzales, TX), and pyrite (Zacatecas, Mexico) were ground with ceramic mortar and sieved in the anaerobic chamber. These were pretreated and washed with 1 M acid solution twice, with 200 mL of 0.1 M dithionite solution to remove oxidized soil mineral surfaces. And all the washed soil minerals were rinsed with ddw several times to remove residual washing solution. These were then freeze-dried, dry-sieved, and stored in the anaerobic chamber.

Magnetite, FeS, and four types of GRs were synthesized. Magnetite was synthesized by adding 0.3 mM $FeCl_2 \cdot 4H_2O$ to the same volume of 0.03 mM $Fe(NO_3)_3 \cdot 9H_2O$ and mixing them with a magnetic stirrer in the reactor with 50 mL/min N_2 gas was flowing [34]. Two M of NaOH was added into the mixed solution until its pH was raised to 7.2 and then air was bubbled into the suspension. When the pH of suspension was constant without adding base solution, the black precipitates were washed with ddw several times to remove residual Fe(II) in the suspension. The synthesized magnetite was freeze-dried, sieved, and stored in the anaerobic chamber. All soil minerals were used for experiments no later than 7 days after pretreatment and synthesis to preclude an aging effect [35].

Four types of GRs were synthesized following Genin group [36-38]'s partial air oxidation method in which the $Fe(OH)_2$ suspension prepared by mixing Fe(II) with NaOH solutions is induced to be partially oxidized by the atmosphere in the presence of appropriate anion (Cl^-, SO_4^{2-}, CO_3^{2-}, and F^-). After synthesis, the suspension of dark blue-green precipitates was washed 2~3 times with DDW and dried in the anaerobic chamber, because GRs would be oxidized during freeze-drying. FeS was synthesized by the method developed by Butler et al. [20]. 1.236M of Na_2S slowly added to the same volume of $FeCl_2{\cdot}4H_2O$ (1.068M) and mixing them with a magnetic stirrer in the anaerobic chamber for 3 days. It was rinsed with DDW four times to remove sulfide and Fe(II) in the suspension through following the steps. To separate solid from the suspension, FeS suspension was centrifuged at 3000rpm for 15min. During the centrifugation, the supernatant was replaced by DDW and the solid was re-suspended. At the last rinsing step, Tris buffer solution instead of DDW was used to keep the pH of suspension constant.

II.2. Description of Experimental Procedures

II.2.1. Abiotic Reductive Dechlorination of Chlorinated Ethylenes by Iron Bearing Soil Minerals and Phyllosilicates

Batch kinetic experiments with amber borosilicate glass vials (nominally 20mL, Kimble) were conducted to study dechlorination kinetics and to identify transformation products. The vials were sealed by three-layer seals (PTFE tape, lead foil, and PTFE-lined rubber septum) with open-top cap to maintain anaerobic condition and to prevent the loss of chlorinated volatile compounds. Pyrite, magnetite, GR_{SO4}, biotite, vermiculite, and montmorillonite were used as an iron-bearing soil minerals and phyllosilicates for reductive degradation of chlorinated ethylenes. Exact amount of each soil minerals were added to the vials filled with a 10 mM $NaHCO_3$ solution (23.6mL) in the anaerobic chamber. The mass ratios of solid to water were 0.084 (pyrite), 0.063 (magnetite), 0.007 (GR_{SO4}), and 0.085 (biotite, vermiculite, and montmorillonite) resulting in different surface area concentration of pyrite (2340 m^2/L), magnetite (3600 m^2/L), GR_{SO4} (604 m^2/L), biotite (163 m^2/L), vermiculite (2230 m^2/L), and montmorillonite (42000 m^2/L). In order to investigate the effect of Fe(II), a stock Fe(II) solution (0.5M $FeSO_4$) was spiked to the soil mineral suspension except GR_{SO4} resulting in

Fe(II) concentration of 42.6 mM in pyrite and magnetite and of 4.28 mM in others. The pH of soil mineral suspensions were adjusted to 7 (magnetite, GR_{SO4}, biotite, vermiculite, and montomorillonite) and 8 (pyrite) by adding 1M acid or base solutions and kept constant during the experiments. The solid suspensions were equilibrated for two days. The average headspace volume in the vial was 0.6 mL, which would allow less than 1.5% of TCE partitioning to the headspace assuming a dimensionless Henry's law constant of 0.359 at 22ºC [39] and no sorption. The methanolic stock solutions (10-50 μL) were spiked into the soil mineral suspensions to obtain 0.19 mM PCE, 0.25 mM TCE, 0.41 mM cis-DCE, and 0.40 mM VC. The vials were rapidly capped, taken out of the anaerobic chamber, mounted on the tumbler, and then completely mixed at 7 rpm at room temperature (22±0.5ºC). All soil mineral samples and controls (bicarbonate buffer + target organics) were prepared in duplicate or triplicate. Target organics and their transformation products in samples were measured at each sampling time.

The effect of sorbed Fe(II) in biotite samples was investigated by spiking 0-200 μL of 0.5 M Fe(II) stock solution to biotite suspensions with mass ratio of 0.085. The suspension was equilibrated for two days and the TCE stock solution was spiked to the suspension resulting in the initial concentration of 0.3 mM. The suspension pH was kept constant at 8.1 by Tris buffer and the controls were prepared following same way as described above. The samples and controls were also prepared in duplicate and the concentration of target organic in aqueous solution was monitored at each sampling time.

To identify oxidation products of GR_{SO4} after reaction with TCE, a batch experiment with 200-mL amber bottle (Kimble) was conducted. The initial concentration of TCE in the GR_{SO4} suspension with bicarbonate buffer was 0.22 M and mass ratio of GR_{SO4} was 0.01. The bottle was shaken in the anaerobic chamber at 180 rpm at room temperature. The oxidation products of GR_{SO4}, Fe(II), and Cl^- concentrations in samples were monitored at each sampling time. All the samples and controls (bicarbonate buffer + target organics) in this study were run in duplicate or triplicates.

II.2.2. Effect of Transition Metal on Degradation of Chlorinated Organic Compounds by Iron-Bearing Soil Minerals

Amber borosilicate glass vials (24.3 ± 0.1 mL) with a three-layered septum system were used as batch reactor for the kinetic experiments on PCE degradation by four types of GRs with Pt. The pH of GRs suspension was initially adjusted to 7.5 using an HCl and NaOH solution, and then Pt was introduced to each suspension so as to generate 0.5, 1.0, and 2.0 mM , respectively. The producing of modified GR suspension by Pt (GR(Pt)) was finally finished by mixing GR and Pt continuously for 10 min with a magnetic stirrer. An aliquot amount (24.2 mL) was taken from the GR as well as 13.9 g/L GR(Pt) suspensions mixed with a magnetic stirrer, with each amount transferred to the glass vials. At each test, PCE stock solution in methanol (596 mM) was freshly prepared. Ten μL of the PCE stock solution was transferred into the GR and GR(Pt) suspensions by a 10 μL gastight syringe (Hamilton) to initiate the reaction. The vials were rapidly and tightly capped, taken out of the chamber, and placed on a tumbler that provided end-over-end rotation at 7 rpm at the room temperature (25 ± 1 ºC). The degradation kinetics of PCE was investigated by monitoring the PCE concentration in aqueous solution at each sampling point. All the samples and controls (without GR and GR(Pt)) were prepared in duplicate and potential loss (i.e., volatilization and sorption) of target compound was investigated in the same way mentioned above.

Kinetic experiment for CT and 1,1,1-TCA degradation were conducted using amber borosilicate glass vials (24 mL) with an open-top cap and three layers as described above, which successfully maintained anaerobic condition and prevented the loss of chlorinated volatile compounds in the previous study [40]. To remove oxygen possibly attached on the reactor wall, the reactors were kept in an anaerobic chamber for more than 1 day. When the vial was mixed by a magnetic stirrer, 23.80 mL FeS suspension (33 g/L) was transferred to each vial. An exact amount of stock transition metal solution was added to FeS suspension resulting in 1, 2.5, 5, and 10 mM of transition metals. To allow chemical equilibrium the suspensions were mixed for 1 h and then transferred to each vial to identify the effect of transition metals on the reductive dechlorination of target compounds. The reaction was initiated by spiking 100 μL of stock target compound solutions resulting in 1 mM of CT and 0.5 mM of 1,1,1-TCA, respectively. Controls filled with Tris buffer and the target compounds were prepared to check the loss of target compounds due to sorption and volatilization during reaction. The vials were mounted on a tumbler that provided end-over-end rotation at 8 rpm. All the samples and controls were prepared in duplicate and the degradation kinetics of CT and 1,1,1-TCA by FeS with and without transition metals were determined by sacrificial sampling and monitoring the concentration of the target compounds at each sampling point.

II.3. Description of Analytical Procedures

II.3.1. Abiotic Reductive Dechlorination of Chlorinated Ethylenes by Iron Bearing Soil Minerals and Phyllosilicates

Hewlett Packard (HP) G1800A GCD system with a DB-VRX column (60 m × 0.25 mm i.d. × 1.8 μm film thickness, J&W Scientific) and a mass spectrometer detector (MS/EID) was used for measuring target organics and their chlorinated transformation products in pyrite, magnetite, and GR_{SO4} samples. The temperatures of injector and detector were 220°C and 240°C, respectively. The oven temperature was programmed to be isothermal at 80 °C for 8min, increased to 160 °C at the rate of 20 °C /min, and then held for 2 min. After centrifugation of batch reactor, 50 μL of supernatant was rapidly withdrawn with a gastight syringe and transferred to an 2mL extraction vial containing 1 mL of extractant (pentane with toluene as an internal standard) for analysis on target organics and chlorinated transformation products in aqueous solution. Supernatant in the reactors was removed and the target organics and chlorinated transformation products sorbed on soil minerals was extracted by adding 10 mL of the extractant to the solids that remained. Both were extracted on an orbital shaker for 30 min at 250 rpm and a 1 μL sample of each extractant was automatically injected into split/splitless injector at a split ratio of 30:1.

For analysis on target organics and their chlorinated transformation products except trans-dichloroethylene (t-DCE), and VC in biotite, vermiculite, and montmorillonite samples, Hewlett Packard (HP) 5890 gas chromatograph (GC) with an electron capture detector (ECD) and a combination of DB-5 column (30 m × 0.25 mm i.d. × 0.25 μm film thickness, J&W Scientific) and DB-5MS column (30 m × 0.25 mm i.d. × 0.25 μm film thickness, J&W Scientific) was used in the same procedure describe above. c-DCE, t-DCE, and VC were analyzed by HP 6890 GC with a DB-VRX column (60 m × 0.25 mm i.d. × 1.8 μm film thickness, J&W Scientific) and a flame ionization detector (FID) following the same procedure as described below, because responses of the ECD for c-DCE, t-DCE, and VC were weak.

Non-chlorinated transformation products in all the soil samples were identified by HP 6890 gas chromatograph with a GS-Alumina column (30 m × 0.53 mm i.d., J&W Scientific) and a flame ionization detector. The temperature of oven was isothermal at 100 °C and the temperatures of injector and detector were both 150 °C. Ten mL supernatant was rapidly transferred by 10 mL gastight syringe to a 20 mL amber vial. The vial was tightly and rapidly sealed and shaken for 1h at 250 rpm to equilibrate the aqueous and gas phases and stood for 2h at room temperature. Gas-phase samples (50-100 μL) were withdrawn from the headspace with a 100 μL gastight syringe (Hamilton) and introduced into the injector at a split ratio of 5:1. The concentration of C_2 hydrocarbons in aqueous solution and c-DCE, t-DCE, and VC in biotite, vermiculite, and montmorillonite samples were calculated by converting of those in headspace using dimensionless Henry's law constants at room temperature (20.4, ethane; 8.7, ethylene; 1.1, acetylene; 0.14, c-DCE; 0.35, t-DCE; 1.01, VC) (42,44,45).

Chloride concentrations were measured by an ion chromatograph (IC, Dionex 500) equipped with AS9-HC column (250 mm × 4 mm i.d., Dionex) and conductivity detector. A 10 mM Na_2CO_3 solution was used as an eluent and the flow rate was 1 mL/min. A sample (2.5 mL) of suspension or supernatant was filtered with 0.2 μm membrane filter (Whatman), with or without dilution, and injected into the column through a 10 μL sample loop.

Fe(II) and total iron in aqueous solution and in solid suspension were measured by using a Hewlett Packard (HP) 8452A Diode-Array Spectrophotometer using the Ferrozine method [43]. Two and fifth millilitre suspension were rapidly withdrawn with a 5 mL gastight syringe (Hamilton) and filtered with a 0.2 μm membrane filter (Whatman). Each 1 mL filtrate was transferred to polypropylene tube (VWR) containing 9 mL ddw. GR_{SO4} suspension (0.5 mL) was transferred to 1.2 M acid solution (9.5 mL) without separation of oxidation products. It was shaken for 5 min to extract iron and a 10% hydroxylamine solution was added to the sample to reduce Fe(III) to Fe(II) for the measurement of total iron.

II.3.2. Effect of Transition Metal on Degradation of Chlorinated Organic Compounds by Iron-Bearing Soil Minerals

Hewlett-Packard 6890 GC equipped with DB-VRX column and electron capture detector was used to determine PCE and its transformation products (TCE and DCEs) in GRs-Pt samples. Also CT and 1,1,1-TCA and their chlorinated daughter products (CF, MC, and 1,1-DCA) in FeS-transition metal samples were measured by Hewlett–Packard 6890 GC equipped with a flame ionization detector and DB-5 column (J&W Co.: 30 m length, 0.32 mm i.d., and 0.25 m thickness). The batch reactors were centrifuged at 2535g for 20 min, and 10 mL of supernatants were transferred to 24 mL borosilicate vials and capped quickly. The vials were shaken for 1h using an orbital shaker at 200 rpm to equilibrate the gas and liquid phases and then allowed to stand for 1h at room temperature (25 °C). Headspace samples (50 μL) were transferred by a gas-tight micro-syringe (100 μL, Hamilton) and injected manually into the injection port of the GC under the following condition: inlet temperature, 250 °C; split ratio, 5:1; column temperature, isothermal at 60 °C; column flow rate, 1.5 mL He/min; detector temperature, 300 °C.

The measurement of non-chlorinated products (methane, ethane, ethylene, and acetylene) was conducted in the same sample preparation procedure as described above. HP 6890 GC equipped with GC-Alumina column and flame ionization detector was used for GRs-Pt sample and GC equipped with a micro-packed column (2 m length, 1/16" OD, 0.75 mm i.d.) containing 80/100 Carboxen-1004 was used for FeS-transition metal sample under the following condition: inlet temperature, 250 °C; column temperature, isothermal at 120 °C; column flowrate, 6 mL He/min; detector temperature, 250 °C; splitless mode.

The amounts of target and transformation products sorbed on soil mineral, septum, and reactor wall were determined by solid phase extraction in which supernatant of centrifuged sample was decanted and replaced by an exact amount of extractant (hexane, HPLC, Sigma-Aldrich). Extraction vial was shaken by orbital shaker at 200 rpm for 2 h. The extractant was analyzed by a GC (HP 5890) with an electron capture detector and HP-5 column (J&W Co.: 30 m length, 0.32 mm i.d., and 0.25 μm thickness).

The morphology of soil mineral surfaces were analyzed by scanning electron microscope (SEM: FEI XL-30 FEG) with energy dispersive spectrometer (EDS). A droplet of diluted soil mineral suspension with and without additives (Pt for GRs and transition metal or HS^- for FeS) was dried on aluminum foil under the anaerobic atmosphere of anaerobic chamber and then coated with Os, respectively.

X-ray photoelectron spectroscopy (XPS:PHI 5800) analysis was conducted to identify the oxidation state of iron, sulfur, Pt, or transition metals on the surface of soil minerals using mono-chromatized Al Kα X-ray (1486.6 eV) with the source power of 15 kV and 24 mA. The Shirley baseline and Gaussian-Lorentzian peak shape was used for the data fitting. A droplet of soil mineral suspension with and without additives (Pt or transition metals and HS^-) was carefully dried on the aluminum foil in the anaerobic chamber. There was no further effort to prevent the oxidation of samples during the XPS analysis.

III. RESULTS AND DISCUSSION

III.1. Abiotic Reductive Dechlorination of Chlorinated Ethylenes by Iron Bearing Soil Minerals and Phyllosilicates

III.1.1. Treatment of Kinetic Data

The rapid disappearance of chlorinated ethylenes in the iron-bearing soil mineral and phyllosilicate suspensions showed initially, followed by slower rates of removal or constant concentrations. Different behavior of chlorinated organics has been observed in most other abiotic reductive transformations, where the kinetics are usually described by a pseudo-first-order or zero-order rate law [18-21,25,41,44]. This behavior obtained in this study can be described by recognizing that the iron-bearing soil minerals and phyllosilicates have a limited reductive capacity and that it was being consumed by reaction with target organics during the experiment. A modified Langmuir-Hinshelwood kinetic model that considers the effect of reductive capacity was used to describe kinetics of reductive dechlorination by iron-bearing soil minerals and phyllosilicates.

$$r_{decay} = kC_{\equiv SCE} = \frac{kC_{RC}C_{CE}}{1/K + C_{CE}} \tag{1}$$

where k is the rate constant for the decay of target chlorinated ethylenes at the reactive sites; $C_{\equiv SCE}$ is the concentration of target chlorinated ethylenes adsorbed on the reactive sites; C_{RC} is the concentration of reductive capacity of soil minerals and phyllosilicates for target chlorinated ethylenes; C_{CE} is the aqueous concentration of target chlorinated ethylenes; and K is the sorption coefficient of target chlorinated ethylenes on reactive sites. The model assumed that 1) target chlorinated ethylenes adsorb onto a finite number of reactive sites on the surfaces of iron-bearing soil minerals and phyllosilicates, 2) reductive dechlorination of the target chlorinated ethylenes occurs at these sites by a first-order reaction resulting in loss of activity of the sites, 3) the reactive sites are the source of reductive capacity of iron-bearing soil minerals and phyllosilicates for target chlorinated ethylenes. The reductive capacity of the iron-bearing soil minerals and phyllosilicates is assumed to be used only for the reductive transformation of the target chlorinated ethylenes, not for any transformation products. Therefore, the concentration of reductive capacity at any time can be calculated as the difference in the initial reductive capacity and the change in total concentration of target chlorinated ethylenes. The total concentration can be obtained by multiplying the aqueous concentration by partitioning factor.

$$C_{RC} = C^0_{RC} - (C^0_{CE,Total} - C_{CE,Total}) = C^0_{RC} - p_{CE}(C^0_{CE} - C_{CE}) \quad (2)$$

$$p_{CE} = (1 + H_{CE}\frac{V_g}{V_{aq}} + k_s) \quad (3)$$

where $C^0{}_{RC}$ is the initial concentration of reductive capacity of iron-bearing soil mineral and phyllosilicates for the target chlorinated ethylene; $C^0{}_{CE,Total}$ is total concentration of target chlorinated ethylene at time equal to zero; $C_{CE,Total}$ is the total concentration of target chlorinated ethylene in all phases; $C^0{}_{CE}$ is the initial aqueous concentration of target chlorinated ethylene; p_{CE} is the equilibrium partitioning factor that equals the ratio of the amount of target chlorinated ethylene in all phases to that in the aqueous phase; H_{CE} is the dimensionless Henry's law constant for target chlorinated ethylene; V_g and V_{aq} are the volumes of gas and aqueous phases; and k_s is the partition coefficient of target chlorinated ethylenes to solid phases that are assumed to be non-reactive. The rate equation can then be described in terms of measured target chlorinated ethylene concentrations as follows.

$$r_{decay} = \frac{k\{C^0_{RC} - p_{CE}(C^0_{CE} - C_{CE})\}C_{CE}}{1/K + C_{CE}} \quad (4)$$

The material balance equation can be combined with the rate equation described above to give the following relationship expressed in terms of aqueous concentrations.

$$\frac{dC_{CE}}{dt} = -\frac{(k/p_{CE})\{C^0_{RC} - p_{CE}(C^0_{CE} - C_{CE})\}C_{CE}}{1/K + C_{CE}} \quad (5)$$

The kinetic parameters (k, $C^0{}_{RC}$, and K) were obtained by an optimization procedure using MATLAB® (MathWorks Inc.). This procedure solves Equation (5) numerically by a fourth-order Runge-Kutta method, calculates the sum of squares for the parameters, and then minimizes it by adjusting values of the parameters with the Marquardt-Levenberg algorithm.

Corrected pseudo-first-order initial rate constants (k_1) were calculated using Equation (6) to compare the dechlorination kinetics reported here to those reported by others, which was derived from Equations (1) and (5).

$$k_1 = \frac{(k/p_{CE})C_{RC}^0}{1/K + C_{CE}^0} \tag{6}$$

k_1 was then normalized by the surface area of the soil minerals to give a surface area-normalized pseudo-first-order initial rate constant ($k_{1,sa}$).

III.1.2. Abiotic Reductive Dechlorination of Chlorinated Ethylenes by Pyrite

Figure 1 shows the degradation of chlorinated ethylenes in controls and pyrite suspensions. TCE concentration in controls abruptly decreased at the start of the experiment due to sorption. It reached approximately 97% of initial

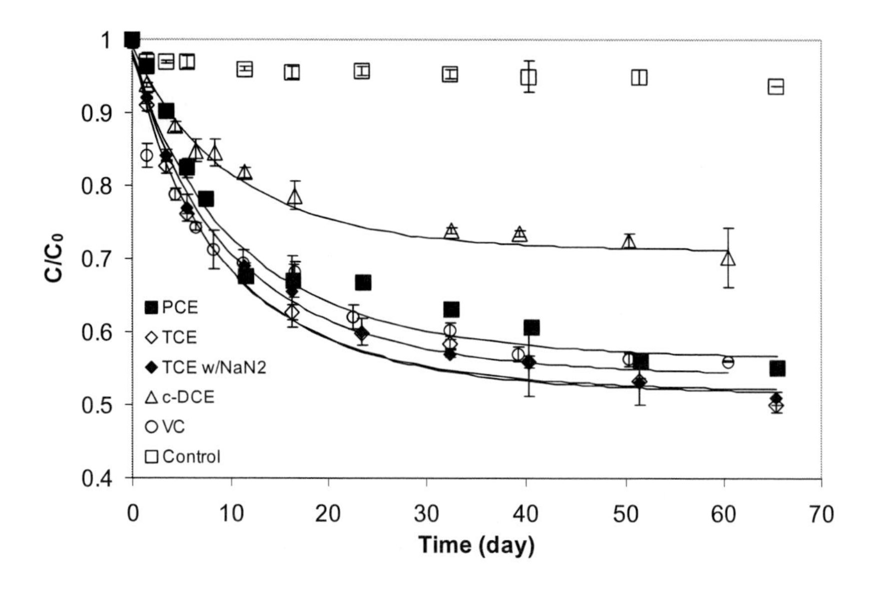

Figure 1. Reductive dechlorination of chlorinated ethylenes in 0.084 g/g pyrite suspensions. Error bars are ranges of relative concentration of target organics. Curves represent kinetic model fits. Control is the control sample for TCE.

TCE concentration at 1.5 day and then gradually decreased over time reaching 94% of the initial value (65.3 days). The concentration of PCE in the control (data not shown) also rapidly dropped and reached equilibrium at 1.5 day (approximately 94% of initial PCE). In cases of c-DCE and VC controls (data not shown), the concentrations of target organics were relatively constant (average > 98%), which indicates that partitioning of c-DCE and VC to the solid phase was negligible.

Chart 1. Kinetic parameters, transformation products and their recoveries, and target organic[a] remaining in pyrite suspensions[b] at the last sampling time

target organics (last sampling time)	specific reductive capacity (μM/g)[c]	K (mM^{-1})	k (day^{-1})	R^{2d}	product recovery and target organic remaining (%)
PCE (65.5 day)	1.01 (±1.9%)[f]	0.642 (±1.5%)	1.01 (±2.2%)	0.974	TCE: 6.7 C_2H_2: 32.0 C_2H_4: 6.0 PCE: 55 total[e]: 99.7
TCE (65.3 day)	1.48 (±2.1%)	0.345 (±3.1%)	1.59 (±1.3%)	0.988	c-DCE: 3.3 C_2H_2: 43.0
				0.989*	C_2H_4: 2.2
	1.47 (±2.4%)*	0.346 (±2.9%)*	1.60 (±1.5%)*		TCE: 50 total[e]: 98.5
c-DCE (60.4 day)	1.49 (±2.4%)	0.300 (±1.8%)	0.984 (±2.1%)	0.983	C_2H_2: 13.2 C_2H_4: 5.2 DCE: 70.2 total[e]: 88.6
VC (60.4 day)	2.26 (±2.6%)	0.187(±3.4 %)	1.71 (±3.5%)	0.970	C_2H_4: 34 C_2H_6: 5.0 VC: 56 total[e]: 95

[a] Initial concentrations of target organics were 0.19 (PCE), 0.25 (TCE), 0.41 (c-DCE), and 0.40 mM (VC). pH of soil mineral suspension was controlled at 8. [b]mass ratio of solid to water was 0.084. [c]specific reductive capacity was obtained by dividing C^0_{RC} by the solid concentration. [d]R^2 values of non-linear regression for kinetic parameters. [e]total is total carbon mass balance for pyrite. [f]uncertainties represent 95% confidence limits. *20 mg NaN_3 was added. **Fe(II) was added. [Fe(II)] = 42.6 mM.

Chart 1 shows the kinetic parameters, transformation products and their recoveries, and target chlorinated ethylene remaining at the last sampling time.

Recovery of transformation product and target chlorinated ethylene remaining were calculated by dividing total concentrations of transformation product and target chlorinated ethylene at the last sampling time by the initial concentration of target chlorinated ethylene. The loss of total recovery of target chlorinated ethylenes and transformation products may be caused by the formation of non-detectable products, inaccuracy in Henry's law constants, and volatilization loss during the sampling procedure.

Approximately, 29.8 - 50% of initial target chlorinated ethylenes were removed by pyrite suspensions and the 88.6 - 99% of carbon mass balance was obtained at the last sampling times. Target organics showed a pseudo-first-order decay during early reaction time, which is similar to that observed in reductive dechlorination of chlorinated ethylenes by meta-stable iron sulfides and Zn(0) [19,21]. However, the degradation rate by pyrite decreased as the reaction proceeded and the reductive capacity of pyrite was consumed. Although the constants are similar to each other differing by a factor of only 1.7, the rate constant for the reductive dechlorination of VC at the reactive surfaces of pyrite has the greatest value followed by those for TCE, PCE, and c-DCE. Over the 70-day sampling period, there were no significant differences observed for the kinetics of TCE removal between the samples with and without biocide (NaN_3), which shows that the reductive dechlorination process in these experiments was abiotic.

Chart 2 shows the surface area-normalized pseudo-first-order initial rate constants. Rate constants for TCE were the highest, followed by those for VC, PCE, and c-DCE. This result is interesting because the calculated reduction potentials for one (or two) electron reduction of PCE and c-DCE are greater than those of TCE and VC, respectively [41,44], which indicates that PCE and c-DCE should be more susceptible to reductive dechlorination. c-DCE has the smallest value, which agrees with previous research [41]. The rate constants for pyrite are smaller than those reported for other reductants. They are approximately two to five orders of magnitude smaller than those reported for the reductive transformation of target organics by Zn(0) [41] and was 1390 - 2900 times smaller than those reported for mackinawite and troilite [19,21]. Microorganisms under Fe(III)-reducing and methanogenic conditions removed 8% and 5% of VC in 37 days [44], which are much smaller than the removals by pyrite in 32 days (40%) reported here. These results indicate that pyrite is less reactive than meta-stable iron sulfides and Zn(0) for reductive dechlorination of chlorinated ethylenes but could more important than microorganisms under some conditions in affecting the fate of chlorinated ethylenes in natural and engineered systems.

Chart 2. Surface area-normalized pseudo-first-order initial rate constants ($k_{1,sa}$)[a] for the reductive dechlorination of chlorinated ethylenes by a variety of reductants

Reductants (surface area concentration: m^2/L)	PCE	TCE	c-DCE	VC	reference
pyrite[b] (2340)	1.97×10^{-5}	2.53×10^{-5}	1.32×10^{-5}	2.27×10^{-5}	this chapter
magnetite[b] (3600)	8.38×10^{-7}	7.21×10^{-7}	5.60×10^{-7}	5.64×10^{-7}	this chapter
			5.74×10^{-6}*	5.78×10^{-6}*	
magnetite[c] (81)		4.56×10^{-4}			(*19*)
Zn(0)[d] (5.47)	8.28	0.075×10^{-2}	8.45×10^{-5}	2.40×10^{-3}	(*41*)
mackinawite[e] (0.5)	2.74×10^{-2}	7.15×10^{-2}			(*21*)
troilite[f] (0.5)		5.52×10^{-2}[g]			(*19*)

[a] Unit of surface area-normalized pseudo-first-order initial rate constant ($Lm^{-2}day^{-1}$). [b]initial target organic concentration: 0.19 (PCE), 0.25 (TCE), 0.41 (c-DCE), and 0.40 mM (VC), pH of soil mineral suspension: 7 (magnetite) and 8 (pyrite). [c]experimental condition is not available. [d]initial target organic concentration: 0.015 to 0.35 mM (PCE, TCE, and c-DCE), 0.003 to 0.03 mM (VC), pH of Zn(0) suspension (50 mM Tris buffer): 7.2. [e]initial target organic concentration: 0.013 (PCE) and 0.015 μM (TCE), pH of mackinawite suspension (50 mM Tris buffer): 8.3. [f]initial concentration is not available, pH: 7.5 (influent) and 7.3 (effluent). [g]the surface area-normalized initial rate constant by troilite was calculated assuming surface area concentration of troilite is 0.5 m^2/L. *Fe(II) was added. [Fe(II)] = 42.6 mM.

III.1.3. Abiotic Reductive Dechlorination of Chlorinated Ethylenes by Magnetite

Figure 2 shows the decay of target chlorinated ethylenes (PCE, TCE, c-DCE and VC) in controls and magnetite suspension with and without addition of 42.6 mM Fe(II).

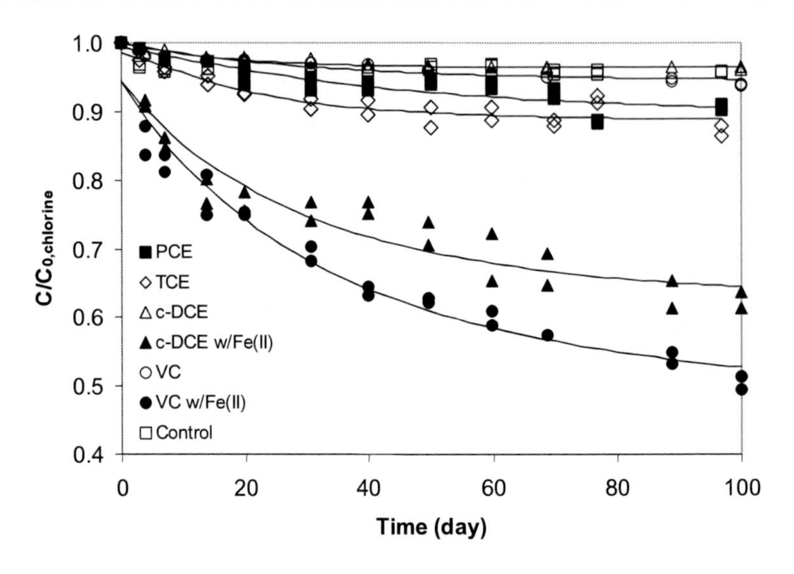

Figure 2. Reductive dechlorination of chlorinated ethylenes in magnetite suspension (0.063 g/g) with and without 42.6 mM Fe(II) addition. Curves represent kinetic model fits. Control is the control sample for PCE.

Concentrations are shown as total chlorine concentration (organic chlorine and chloride) relative to the initial total chlorine concentration. The concentration of PCE in controls rapidly dropped due to the partitioning to the solid phase and reached equilibrium concentration (approximately 96% of initial PCE concentration) in 3 days as observed in pyrite suspension. The concentrations of controls for TCE, c-DCE, and VC (not shown) also decreased rapidly and reached equilibrium (96.5, 97.7, and 98%, respectively) in 3 days. The amount of sorption of each target chlorinated ethylene in magnetite suspension was greater than that in pyrite suspension. 2.1 times greater surface area of magnetite than that of pyrite may cause higher sorption capacity of magnetite for target chlorinated ethylenes.

Approximately, 3.9 to 13.7% of the initial target chlorinated ethylenes were removed in magnetite suspension and 1.95 to 10.7% were transformed to chloride in 100 days. The relative total chlorine concentration of target chlorinated ethylenes (chlorine balance) in magnetite suspension was 95.6 to 98.1% at the last sampling time (Chart 3). Chart 3 shows that the rate constant for the reductive dechlorination of TCE at the reactive magnetite surfaces is the greatest, followed by those for PCE, VC, and c-DCE. The specific reductive capacity of magnetite for target chlorinated ethylenes was 2.8 to 4 times smaller than that of pyrite.

Chart 3. Kinetic parameters, transformation products and their recoveries, and target organic[a] remaining in magnetite suspensions[b] at the last sampling time

target organics (last sampling time)	specific reductive capacity (μM/g)[c]	K (mM^{-1})	k (day^{-1})	R^{2d}	product recovery and target organic remaining (%)
PCE (96.8 day)	0.33 (±7.4%)	0.700 (±3.1%)	0.202 (±8.4%)	0.910	chloride: 5.5 PCE: 90.1 total[e]: 95.6
TCE (96.8 day)	0.37 (±5.6%)	0.503 (±2.8%)	0.254 (±4.1%)	0.961	chloride: 10.7 TCE: 86.3 total[e]: 97
c-DCE (99.9 day)	0.54 (±10.3%)	0.501 (±11.2%)	0.185 (±12.5%)	0.817 0.872**	chloride: 1.95/28.7** DCE: 96.1/61.3** total[e]: 98.1/90**
	2.06 (±7.6%)**	0.344 (±5.3%)**	0.28 (±8.08%)**		
VC (99.9 day)	0.73 (±9.7%)	0.3 (±9.3%)	0.193 (±10.8%)	0.823 0.943**	chloride: 4.2/43.9** VC: 93.7/48.3** total[e]: 97.9/92.2**
	2.86 (±3.3%)**	0.2 (±4.9%)**	0.35 (±5.3%)**		

[a] Initial concentrations of target organics were 0.19 (PCE), 0.25 (TCE), 0.41 (c-DCE), and 0.40 mM (VC). pH of soil mineral suspension was controlled at 7. [b]mass ratio of solid to water was 0.063. [c]specific reductive capacity was obtained by dividing C^0_{RC} by the solid concentration. [d]R^2 values of non-linear regression for kinetic parameters. [e]total is total carbon mass balance for magnetite. [f]uncertainties represent 95% confidence limits. *20 mg NaN_3 was added. **Fe(II) was added. [Fe(II)] = 42.6 mM.

The sorption coefficients of target organics in magnetite suspension were slightly greater than those in pyrite suspension by a factor of 1.1 to 1.7 [25].

The $k_{1,sa}$ for the reductive dechlorination of PCE by magnetite (Chart 2) has the greatest value followed by those for TCE, VC, and c-DCE. The reductive dechlorination of target chlorinated ethylenes by magnetite is much slower than dechlorination by other reductants. For example, values of $k_{1,sa}$ for magnetite were approximately 23.5 to 40.3 times smaller than those for pyrite and 2 to 7 orders of magnitude smaller than those for Zn(0) [41]. The $k_{1,sa}$ for reductive dechlorination of TCE by magnetite observed in this study was 630 times smaller than that reported by Sivavec et al [19,46]. Ferrous iron was added to simulate the condition in which microbial reduction of Fe(III) to Fe(II) would provide a continuous source of reducing power. No significant difference was observed on the decay of c-DCE and VC between the controls with and without Fe(II) addition (Data not shown) indicating that the reductive dechlorination of target organics by dissolved Fe(II) is not important. The addition of Fe(II) to magnetite suspensions increased the dechlorination rates of c-DCE and VC by a factor of 10 and increased the specific reductive capacity of magnetite for c-DCE and VC by a factor of approximately 4. Similar results were also observed in the reductive dechlorination of c-DCE in magnetite suspension with and without Fe(II) addition. The addition of Fe(II) increased the reactivity of magnetite by providing continuous reductant source. The enhanced reactivity of magnetite due to Fe(II) addition indicates the potential link between biotic and abiotic processes in the reductive dechlorination of c-DCE and VC in contaminated plumes. An enzyme exuded from microorganisms (e.g., *Geobacter metallireducens*) can reduce hydrous ferric oxide to magnetite [47] and/or to Fe(II) surface species sorbed onto the magnetite under Fe(III) reducing conditions. These biogenic magnetite and Fe(II) surface species could be regenerated by microbial enzymes after their oxidation by chlorinated organics. The effect of Fe(II) addition to the magnetite suspension, therefore, may be due to the reductive regeneration of magnetite by Fe(II) or by the activity of Fe(II) bound to magnetite [22].

III.1.4. Abiotic Reductive Dechlorination of Chlorinated Ethylenes by Green Rust

The decay of target chlorinated ethylenes in controls and GR_{SO4} suspension is represented in Figure 3. PCE concentration in controls rapidly decreased due to the partitioning to the solid phases and reached equilibrium concentration (91.2% of initial PCE concentration) at the first sampling point (1.6 day) as observed in pyrite and magnetite suspensions. The concentrations of TCE, c-DCE, and VC (not shown) in controls also dropped rapidly and reached equilibrium (95.9, 97, and 98% of initial target chlorinated ethylene concentrations, respectively) at 1.6 day. The rate constants for the dechlorination of target chlorinated ethylene in GR_{SO4} suspension (Chart 4) are similar to those observed in pyrite suspension [4]. The rate constant for the dechlorination of PCE at the reactive surfaces of GR_{SO4} was greatest followed by those of VC, TCE, and c-DCE.

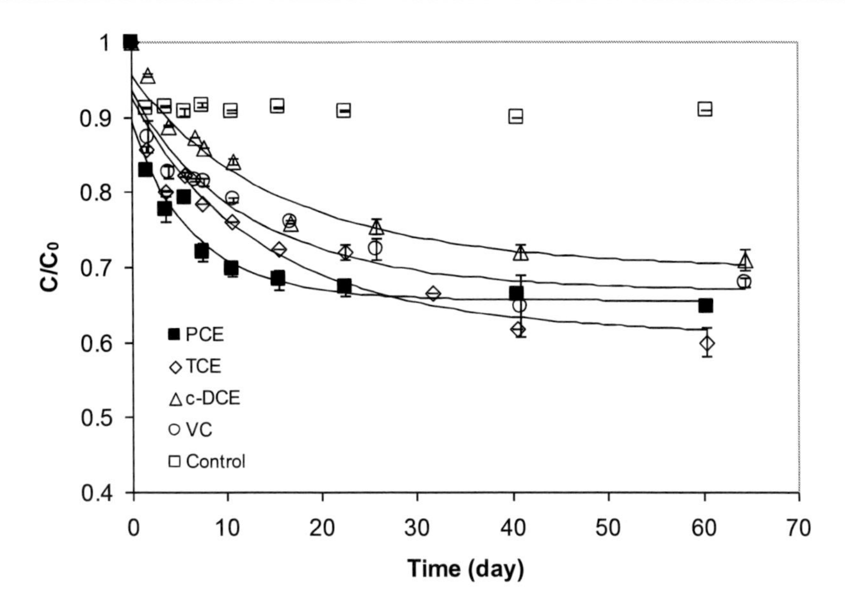

Figure 3. Reductive dechlorination of chlorinated ethylenes in GR_{SO4} suspension (0.007 g/g). Error bars are ranges of relative concentration of target organics. Curves represent kinetic model fits. Control is the control sample for PCE.

The rate constant for the dechlorination of c-DCE by GR_{SO4} was the smallest among the target organics, which is similar to results that have been reported for the reductive dechlorination of chlorinated ethylenes by pyrite, magnetite, and Zn(0) [41,48]. The corrected pseudo-first-order initial rate constants for target organics normalized by GR_{SO4} surface area (Chart 4) were 3.4 to 8.2 times greater than those in pyrite suspension [48]. Approximately, 29 to 40% of initial target chlorinated ethylenes were removed in GR_{SO4} suspension in 70 days and total amount of carbon recovered at the last sampling time was 70 to 90.8% of the initial amount. The loss of total carbon recovery may be caused by the formation of non-detectable products, inaccuracy in Henry's law constants, and volatilization loss during the sampling procedure as previously observed in pyrite and magnetite suspensions.

The main PCE transformation products were acetylene and ethylene, which accounted for 14.5 and 2.0% of the removed PCE, respectively. In contrast to the microbial hydrogenolysis of PCE [17,49] and the reductive transformation of PCE by Vitamin B_{12} with Ti(III)-citrate [44,50], no chlorinated daughter products were observed during the reaction. No further reduction of ethylene to ethane was observed either.

Chart 4. Kinetic parameters, transformation products and their recoveries, and target organic remaining in GR_{SO4} suspensions at the last sampling time

target organics (last sampling time)	S_R (μM/g)	K (mM^{-1})	k (day^{-1})	R^2 [a]	$k_{1,sa}$ ($Lm^{-2}day^{-1}$)[b]	product recovery and target organic remaining (%)
PCE (60.4 day)	9.86 (±10.1%)[d]	1.22 (±4.3%)	1.59 (±6.3%)	0.911	1.62×10^{-4}	C_2H_2: 5.1 C_2H_4: 0.7 PCE: 64.7 total[c]: 70.5
TCE (60.4 day)	14.4 (±13.2%)	0.76 (±4.4%)	0.90 (±8.6%)	0.841	8.55×10^{-5}	C_2H_2: 8.8 C_2H_4: 1.2 TCE: 60 total[c]: 70
c-DCE (64.4 day)	16.9 (±6.0%)	0.633 (±2.0%)	0.592 (±4.4%)	0.937	5.19×10^{-5}	C_2H_2: 0.8 C_2H_6: 8.5 C_2H_4: 5.0 c-DCE: 71 total[c]: 85.3
VC (64.4 day)	18.0 (±4.3%)	0.53 (±2.4%)	0.94 (±2.8%)	0.970	7.77×10^{-5}	C_2H_4: 20.3 C_2H_6: 2.5 VC: 68 total[c]: 90.8

[a] R^2 values of non-linear regression for kinetic parameters. [b]surface area-normalized pseudo-first-order initial rate constant for reductive dechlorination of chlorinated ethylenes, surface area concentration of GR_{SO4} was $604 m^2/L$. [c]total is total carbon mass balance in GR_{SO4} suspension. [d]uncertainties represent 95% confidence limits.

III.1.5. Abiotic Reductive Dechlorination of Chlorinated Ethylenes by Iron Bearing Phyllosilicate (Biotite, Vermiculite, and Momtmorillonite)

The concentrations of target chlorinated ethylenes in controls with and without Fe(II) addition rapidly decreased and reached constant concentrations (95 to 98% and 96 to 99% of initial target organics, respectively) after the first sampling time (3.7 days for PCE and TCE and 2.5 days for *c*-DCE and VC) as observed in pyrite, magnetite, and GR_{SO4}. In contrast to the solid phase partitioning coefficients (k_s) of chlorinated ethylenes in pyrite, magnetite, and GR_{SO4} suspensions, the partitioning coefficients in the phyllosilicate suspensions were 1.2 to 3.4 times greater than those in controls without phyllosilicates, which indicates that a substantial amount (16.7 to 70.6%) of target chlorinated ethylene from solution were partitioned to the iron-bearing phyllosilicate surfaces. Partitioning of most target chlorinated ethylenes to montmorillonite was greater than that to biotite and vermiculite by one to two orders of magnitude, which was probably due to the greater surface area of montmorillonite.

Figure 4 shows the reductive dechlorination of PCE in iron-bearing phyllosilicate suspensions with and without Fe(II) addition. Chart 5 represents values of the parameters used by the kinetic model, which was obtained during the dechlorination of other chlorinated ethylenes as described above. The rate constants for the reductive dechlorination of target chlorinated ethylenes at the reactive biotite surfaces both with and without Fe(II) addition were 1.5 to 5 times greater than those at the surfaces of vermiculite and montmorillonite. This result may be due to the higher content of sites on biotite that contain Fe(II). The Fe(II) content of biotite is 8 and 97.5 times higher than that of vermiculite and montmorillonite, respectively [40,48]. The rate constants of target chlorinated ethylenes by iron-bearing phyllosilicate significantly increased by addition of Fe(II). This enhanced reactivity may be caused by the regeneration of active sites on the phyllosilicates resulting from reaction with Fe(II) or by the reactivity of Fe(II) that binds to the phyllosilicate surfaces.

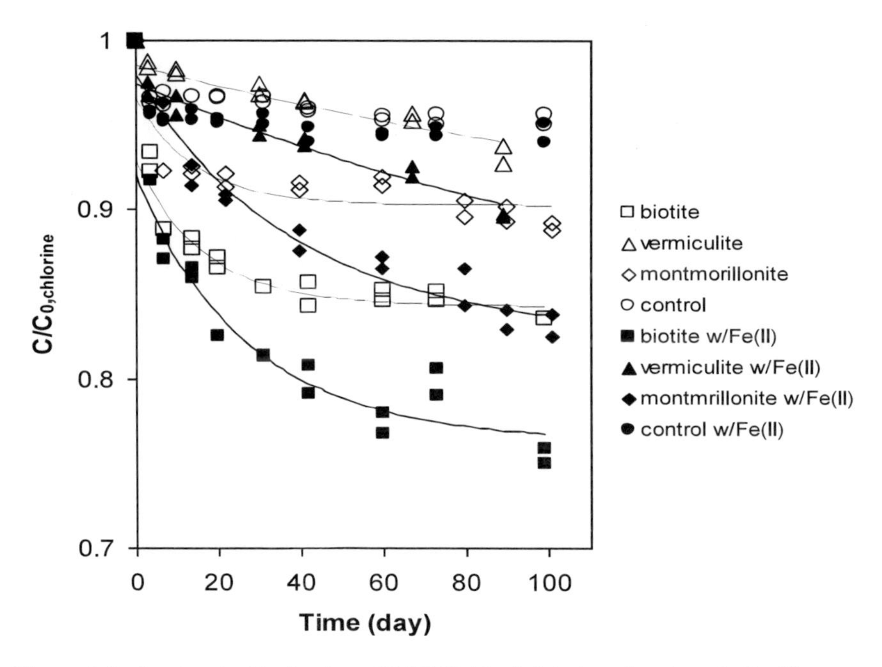

Figure 4. Reductive dechlorination of PCE (0.19 mM) in iron-bearing phyllosilicate suspensions (0.085 g g^{-1}) with and without Fe(II) addition (4.28 mM). Curves represent kinetic model predictions based on the parameters in Chart 5.

Chart 5. Kinetic parameters, recoveries of chloride and total chlorine, and percentage of target organic remaining for the reductive dechlorination of chlorinated ethylenes[a] by iron-bearing phyllosilicates[b] with and without Fe(II) addition[c] at the last sampling time

soil minerals	target organics (last sampling time)	S_R (μM g^{-1})	K (mM^{-1})	k (day^{-1})	product recovery and target organic remaining (%)
biotite	PCE (98.8 day)	0.400 (±5.3%)[d] 0.565 (±3.7%)*	1.6 (±6.1%) 0.802 (±2.8%)*	0.401 (±8.1%) 0.601 (±5.1%)*	chloride: 4.1 / 16.0* PCE: 83.6 / 75.8* total[e]: 87.7 / 91.8*
	TCE (98.8 day)	0.482 (±7%) 0.647 (±4.0%)*	1.0 (±3.5%) 0.615 (±2.7%)*	0.4 (±5.7%) 0.572(±3.9%)*	chloride: 3.8 / 12.4* TCE: 84.1 / 78.6* total[e]: 87.9 / 91.0*
	c-DCE (98.6 day)	0.612 (±4.2%) 0.953 (±3.1%)*	0.904 (±3.9%) 0.380 (±4.1%)*	0.325 (±4.2%) 0.482 (±4.8%)*	chloride: 3.9 / 10.8* *c*-DCE: 85.6 / 80.0* total[e]: 89.5 / 90.8*
	VC (98.6 day)	1.06 (±7.1%) 1.73 (±5.1%)*	0.597 (±5.3%) 0.2 (±2.8%)*	0.358 (±6%) 0.555 (±3.7%)*	chloride: 3.0 / 17.6* VC: 86.2 / 74.4* total[e]: 89.2 / 92.0*
Vermicu -lite	PCE (89 day)	0.177 (±6.1%) 0.282 (±10.1%)*	1.4 (±7.2%) 1.0 (±9.8%)*	0.08 (±10.4%) 0.125 (±15.2%)*	chloride: 2.0 / 5.0* PCE: 92.8 / 89.6* total[e]: 94.8 / 94.6*
	TCE (89 day)	0.188 (±7.9%) 0.294 (±10.7%)*	1.1 (±9.4%) 0.803 (±6.8%)*	0.08 (±16.2%) 0.165 (±11.5%)*	chloride: 2.4 / 5* TCE: 94.5 / 90.6* total[e]: 96.9 / 95.6*
	c-DCE (89 day)	0.635 (±2.7%) 0.871 (±2.5%)*	0.775 (±3.2%) 0.4 (±3.4%)*	0.162 (±4.1%) 0.4 (±2.9%)*	chloride: 1.8 / 4.5* *c*-DCE: 89.5 / 84.4* total[e]: 91.3 / 88.9*
	VC (89 day)	1.01 (±11.4%) 1.41 (±12.4%)*	0.502 (±6.3%) 0.506(±12.1 %)*	0.355 (±12.6%) 0.537 (±20.4%)*	chloride: 1.4 / 3.0* VC: 87.9 / 83.9* total[e]: 89.3 / 86.9*
Montmo -rillonite	PCE (100.6 day)	0.271 (±7.3%) 0.388 (±5.1%)*	1.6 (±6.3%) 0.685 (±4.0%)*	0.2 (±7.4%) 0.396 (±4.7%)*	chloride: 3.8 / 9.7* PCE: 88.8 / 80.8* total[e]: 92.6 / 93.6*

soil minerals	target organics (last sampling time)	S_R (μM g^{-1})	K (mM^{-1})	k (day^{-1})	product recovery and target organic remaining (%)
	TCE (100.6 day)	0.282 (±10.1%) 0.577 (±7.1%)*	1.0 (±8.7%) 0.6 (±5.2%)*	0.306 (±17.4%) 0.5 (±7.4%)*	chloride: 3.5 / 10.1* TCE: 90.5 / 82.0* total[e]: 94 / 92.1*
	c-DCE (100.6 day)	0.318 (±5.2%) 0.729 (±7.5%)*	0.904 (±4.7%) 0.4 (±8.2%)*	0.171 (±5.1%) 0.3 (±8.7%)*	chloride: 1.2 / 4.5* *c*-DCE: 94.3 / 86.7* total[e]: 95.5 / 91.2*
	VC (100.6 day)	0.353 (±8.2%) 1.25 (±6.8%)*	0.502 (±7.6%) 0.301 (±8.2%)*	0.247 (±9.7%) 0.323 (±12%)*	chloride: 0.7 / 5.6* VC: 96 / 85* total[e]: 96.7 / 90.6*

[a] Initial concentrations of target organics were 0.19 (PCE), 0.25 (TCE), 0.41 (*c*-DCE), and 0.64 mM (VC). pH of soil mineral suspension was controlled at 7.0. [b] Mass ratios of solid to water were 0.085 g g^{-1}. [c] Fe(II) concentration: 4.28 mM. [d] Uncertainties represent 95% confidence intervals. The uncertainties are calculated by multiplying standard errors of estimated kinetic parameters by *t*-values. [e] Total is recovery of total chlorine in the phyllosilicate suspension with and without Fe(II) at the last sampling time. * Samples with Fe(II) addition.

The rate constants for phyllosilicates were 1.8 to 20 times smaller than those for pyrite and GR_{SO4} in this chapter. The difference of rate constants among soil minerals may be due to different intrinsic reactivity of active sites or to different concentrations of active sites on each soil mineral. The added Fe(II) is expected to be a part of all active sites and the Fe(II) contents of iron-bearing phyllosilicates are 4 to 400 times lower than those of GR_{SO4}. Although the addition of Fe(II) to iron-bearing phyllosilicate suspensions increased the dechlorination rates, the rate constants with Fe(II) addition were still 1.2 to 12.7 times smaller than those for pyrite and GR_{SO4}. Approximately, 8.8 to 15.3% and 3.7 to 12% of initial *c*-DCE (0.41 mM) and VC (0.64 mM) were removed by abiotic reductive dechlorination of by iron-bearing phyllosilicates with and without Fe(II) addition, respectively, over a 30 day period. Compared to abiotic degradation of target chlorinated ethylenes in this chapter, biological degradation of 1,2-dichloroethylene (1.4 - 80 μM) and VC (57 μM) by Fe(III)-reducing and methanogenic bacteria has been reported as removing 4 to 14% of initial target organics in 37 days [45]. This indicates that the extent of removal of *c*-DCE and VC by iron-bearing phyllosilicates can be similar to those by microorganisms, even when the initial concentrations are higher in the abiotic system. This result implys that abiotic reductive dechlorination of *c*-DCE and VC by iron-bearing phyllosilicates may be important processes that act in combination with biodegradation during natural attenuation of contaminated groundwater.

III.2. Effect of Transition Metal on Degradation of Chlorinated Organic Compounds by Iron-Bearing Soil Minerals

III.2.1. Dechlorination of Chlorinated Compounds by FeS with Transition Metal

Figure 5(a) is the reductive dechlorination of CT by FeS at pH 7.5 following a first-order rate law. 90% of 1 mM CT was reductively degraded in 3h and CF concentration was gradually increased to 0.4 mM. CF concentration was slightly decreased at 2h and then it remained constant. Less chlorinated compounds (e.g., MC) were not observed in this research. CH_4, C_2H_4, and C_2H_6 that non-chlorinated C_1 and C_2 hydrocarbons were not detected either while they were possibly produced during the reductive dechlorination of CT [51]. Adsorption of the target compound and transformation products on the solid surface were also investigated, but no detectable amounts of target and products were recovered from the solid samples during the reaction. The total carbon mass balance was about 50% at the end of reaction based on the measurement of CT and its products. It has been reported by many groups that CT was mainly transformed to CF through hydrogenolysis pathway by iron-bearing soil minerals and the extent of CF production was dependent on the type of soil minerals [31,52-54]. In the Zwank et al study, the portions of CF formation during the reductive dechlorination of 10 μM CT by mackinawite, goethite, magnetite, lepidocrocite, hematite, and siderite were 55%, 33%, 80%, 14%, 17%, and 33%, respectively [54]. Some researchers suggested different reaction pathways in which CT could reductively transformed to CS_2, CO_2, CO and $HCOO^-$ [54,55]. As a result, the obtained low total carbon mass balance in this study may be due to the formation of other transformation products (e.g., CS_2, CO_2, CO and $HCOO^-$), which were not analyzed during the reaction. Figure 5(b) shows the effect of transition metals (Cu(II) and NI(II)) on the degradation kinetics of CT by FeS. The reaction rate constants of FeS with Cu(II) and Ni(II) were 1.59 (±0.357) and 1.32 (±0.363)h^{-1}, respectively, and these were slightly higher than that without a transition metal (1.24 (±0.266)h^{-1}). At the last sampling time, almost 40% of initial CT was transformed to CF by FeS with Cu(II) and Ni(II). Chlorinated methanes and chlorinated products were not measured from the samples during the reaction.

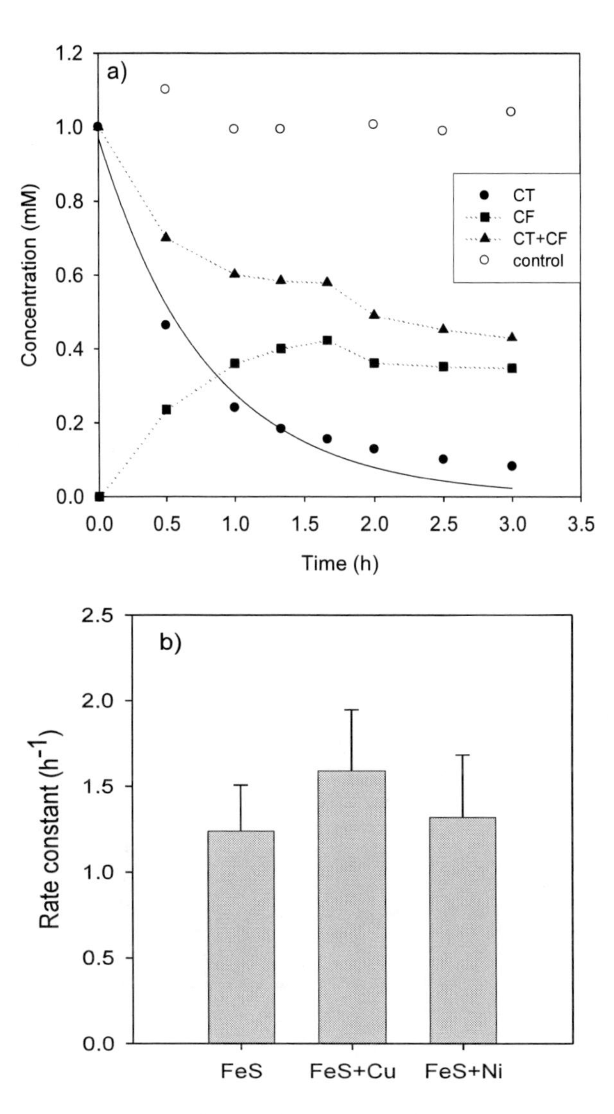

Figure 5. (a) Reductive degradation of CT and formation of CF and (b) kinetic rate constants for the reductive degradation of CT by FeS with 1 mM Cu(II) and Ni(II). FeS suspension was 33 g/L and pH was kept constant at pH 7.5. The initial concentration of CT was 1.0 mM. A solid line represent pseudo-first order model and dot lines are plotted (not fit) to guide the eye and error bars represent 95% confidence intervals of rate constants.

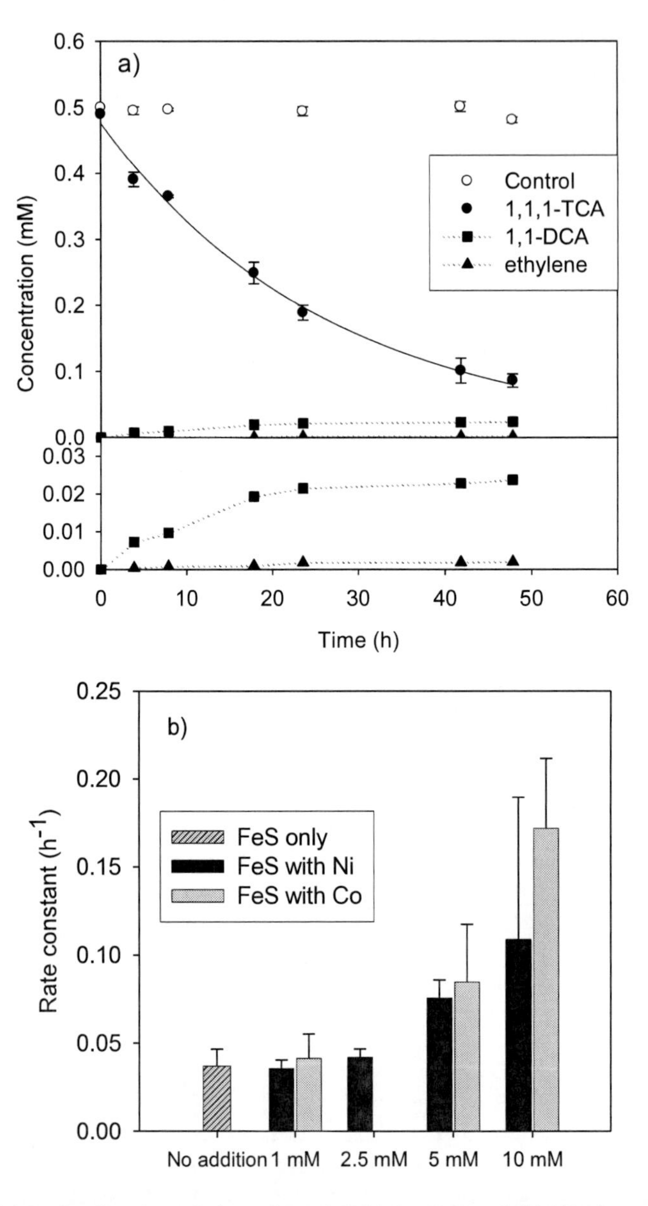

Figure 6. (a) Reductive degradation of 1,1,1-TCA by FeS and (b) kinetic rate constants for the reductive degradation of 1,1,1-TCA by FeS with a trace metal (Ni and Co). FeS suspension was 33 g/L and pH was kept constant at pH 7.5. The initial concentration of 1,1,1-TCA was 0.5 mM. A solid line represent pseudo-first order model and dot lines are plotted (not fit) to guide the eye and error bars represent 95% confidence intervals of rate constants.

The reductive dechlorination of 1,1,1-TCA by FeS was examined in this study (Figure 6(a)). The degradation rate of 1,1,1-TCA was lower than that of CT. Approximately, 82% of initial 1,1,1-TCA (0.5mM) was removed by FeS within 48h. The obtained rate constant for the reductive dechlorination of 1,1,1-TCA by first-order linear fit was 0.0375 (±0.0018)h^{-1}. The formation of 1,1-DCA was observed in the concentration range of 0.003-0.02 mM (~4% of total removal) showing that 1,1,1-TCA was reductively transformed into 1,1,-DCA via hydrogenolysis pathway [26]. Less than 1% of ethylene in total removal was also detected during the experiment. Detectable amounts of 1,1,1-TCA and transformation products were not recovered from the solid surfaces. At the end of sampling time, the total carbon mass balance for the reductive dechlorination of 1,1,1-TCA by FeS was 21% (i.e., 17% of 1,1,1-TCA+4% of 1,1-DCA+0.2% of ethylene). It was significantly lower than that of CT (50%) and this may be due to the formation of other possible transformation products, which were not analyzed from the samples during the reaction. Gander et al. reported a similar result in which 1,1-DCA (2%) and 2-butune (4%) were produced while more than 95% of 175 μM 1,1,1-TCA was removed by FeS [56]. Fennelly and Roberts proposed the reaction pathway of 1,1,1-TCA that 1,1,1-TCA can be initially transformed to 1,1-DCA through hydrogenolysis,1,1-dichloroethylene through dehydrochlorination, and 2,2,3,3-tetrachlorobutane by coupling two dichloroethyl radicals ($H_3C-\dot{C}-Cl_2$) produced by one election reduction to 1,1,1-TCA [26]. Additionally, it has been reported that carbine intermediates ($H_3C-\ddot{C}-Cl$) produced by two-electron reduction to 1,1,1-TCA triggered the formation of acetaldehyde and vinyl chloride [57].

Figure 6(b) represents the reductive dechlorination of 1,1,1-TCA by FeS with Cu(II), Co(II), and Ni(II) from 1 to 10 mM. The reaction rate constants increased with respect to the concentrations of Co(II) and Ni(II) increased, except for the addition of Cu(II). It shows the good linearity between the rate constants and transition metal contents. The slopes for the reaction rate constants of Ni(II) (0.0085 (R^2 = 0.98)) was 5.0 times more reactive than that of Co(II) (0.0017 (R^2 = 0.90)). The increase of Cu(II) concentration in the suspension was not effective to enhance the reductive dechlorination of 1,1,1-TCA by FeS because the mild hydrogenation catalyst such as Cu(II) (IB metal) does not readily donate electron compared to VIIIA metals such as Co(II) and Ni(II) [58]. As main transformation products, 1,1-DCA and ethylene were observed in the reductive dechlorination of 1,1,1-TCA by FeS with Ni(II) and Co(II). At the last sampling time, less than 5% of initial 1,1,1-TCA concentration was observed and it is very similar to that observed in the reductive dechlorination of HCA by FeS with a transition metal from 0.1 to 10 mM [58]. It has been proposed that addition of transition metals can enhance reductive dechlorination rates by the formation of iron-transition metal sulfide from co-precipitation or metal isomorphic substitution during the reaction [58].

On the contrary, it has been reported that transition metals added into ZVI and green rust (GR) suspensions sequentially enhance the reductive dechlorination rates by playing a catalytic role as an electron mediator. Divalent transition metal cations added into GR and ZVI were reductively transformed to zero-valent chemical species. The attached zero-valent chemical species on the surfaces of GR and ZVI facilitated the electron transfer to target compounds. [26-28,31,59].

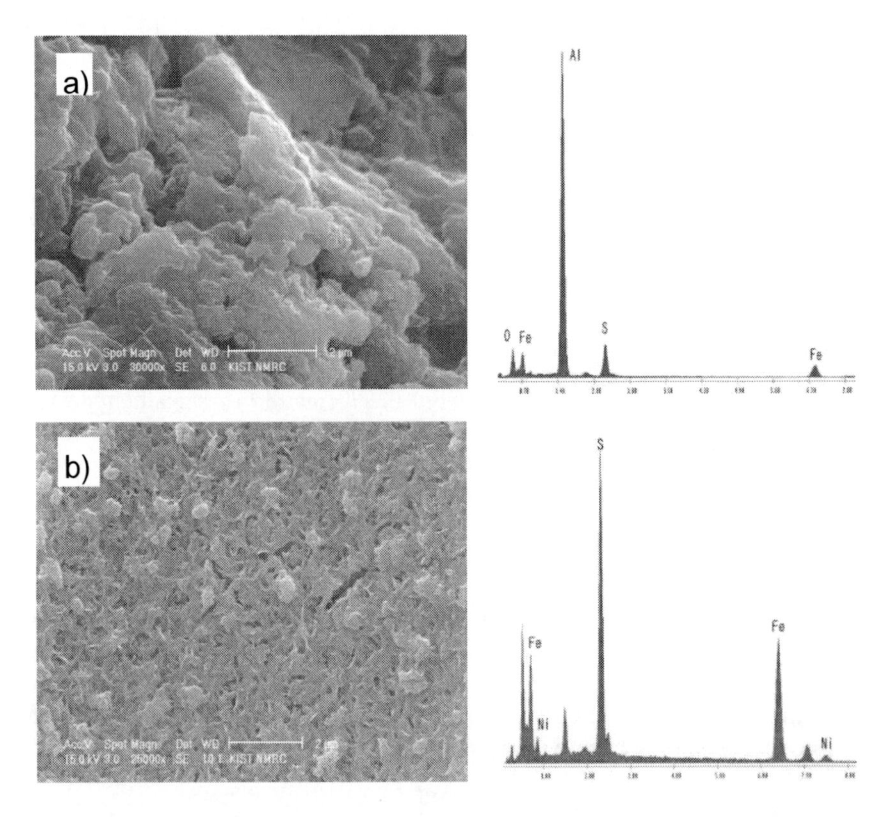

Figure 7. SEM and EDS analysis, (a) the surface of FeS and (b) the surface of FeS with Ni(II).

Figure 7 indicates SEM images (left side) and EDS spectrum (right side) of FeS surfaces without (a) and with a transition metal (Ni(II)) (b). SEM image of FeS shown in Figure 7(a) demonstrates that the amorphous and disordered FeS particles were irregularly aggregated [60]. Any specific shapes of bulk FeS solids were not observed and EDS analysis showed these particles were composed of iron (52.6%) and sulfur (47.4%). The aluminum peak in Figure 7(a) was derived from the aluminum foil used for the drying of diluted FeS suspension. After addition of Ni(II), the shape of solids and surfaces of FeS were significantly changed as shown in Figure 7(b). Figure 7(b) represents that the amassed FeS particles formed big bulk solids and the surface of bulk solids has small cracks. From the EDS analysis, the chemical composition of the bulk was iron (46.9%), sulfur (48.4%), and nickel (4.7%), respectively. The decrease of iron content is possibly due to the isomorphic substitution with the added Ni(II).

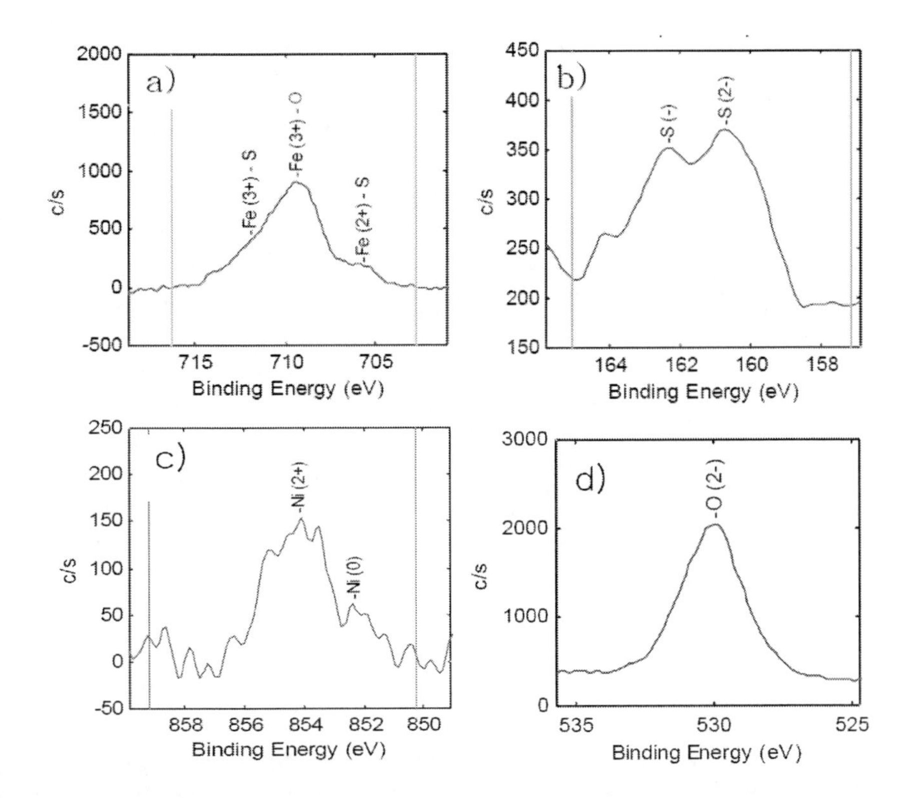

Figure 8. XPS analysis of FeS surface with Ni(II): narrow region scans of (a) Fe(2p3/2), (b) S(2p3/2), (c) Ni(2p3/2), and (d) O(1s).

The oxidation states analysis of Fe, S, and Ni were conducted using XPS and those results were shown in Figure 8. This depicts the narrow scans of Fe(2p3/2), S(2p3/2), O(1s), and Ni(2p3/2) spectra of FeS with Ni(II). By the Shirley baseline and Gaussian–Lorentzian peak shape, each raw spectrum of Fe, S, O, and Ni was smoothed and fitted. There are three peaks in Fe(2p3/2) spectrum consists of 706.0 eV (Fe(II)–S), 709.0 eV (Fe(III)–S), and 711.9 eV (Fe(III)–O), while S(2p3/2) spectrum consists of two peaks at 160.2 eV (S^{2-}) and 162.1 eV (S^{-}), respectively [61,62]. In determining the identity of peaks around 711 eV, Herbert et al. [63] reported as Fe(III)–S, while Thomas et al. [61] identified them as Fe(III)–O. In this experiment, the obtained peak at 711.9 eV identified as Fe(III)–O, which can be supported by the presence of O^{2-} shown in Figure 8(d). The surface chemical composition of FeS were 47.6% Fe(II)–S, 34.4% Fe(III)–S, and 17.9% Fe(III)–O, while the area percentage of S^{2-} and S^{-} in S(2p3/2) spectra (Figure 8(b)) was 86.9% and 13.1%, respectively. The similar area percentage of each target chemical composition (Fe and S) on mackinawite was reported by Mullet et al. [63]. On the surface of FeS, the area percentage of reduced chemical species (Fe(II)–S and S^{2-}) after the addition of Ni(II) was 11.9% and 64%, respectively. It was relatively lower than that without Ni(II) which indicates that Fe(II)–S and S^{2-} on the surfaces of FeS were oxidized during the reduction of Ni(II). Figure 8(c) is the Ni(2p3/2) spectrum consists of two identical peaks at 852.6 and 854.3 eV. The peak at 852.6 eV indicates zero-valent Ni(0) and its area percentage is about 25%, while the peak at 854.3 eV indicates Ni(II) and its area percentage was 75% [64].

Based on SEM and XPS results, it can be proposed that added Ni(II) was bound on the FeS surface and reductively transformed to Ni(0) by the reduced iron and sulfur species on FeS surfaces. Ni(0) increased the reaction rates of reductive dechlorination by facilitating the electron transfer reaction. It has been reported that the metal type of Ni on the surface of reactive solids be able to increase the reaction rates of reductive dechlorination [26,29]. It can reduce chlorinated organic compounds with similar kinetic as ZVI has [30]. In addition, based on the previous studies, we could propose the role of trace metal on the reactive solid surfaces in the reductive dechlorination of target compounds can be explained by a couple of possible reaction mechanisms. Fe(II) in the FeS structure could be replaced by Ni(II) adsorbed on the surfaces of FeS and/or Ni(II) was co-precipitated with FeS resulting in the formation of metal-substituted FeS, such as $Fe_{1-x}Ni_xS$ [65]. The formation of metal-substitute FeS can significantly enhance the reaction kinetics for the reductive dechlorination of chlorinated aliphatic by FeS with a transition metal [58]. NiS can be formed by reaction between Ni(II) and S^{2-} [66] resulting in the increase of reductive dechlorination rates by one order of magnitude compared to that by FeS alone [58].

III.2.2. Dechlorination of PCE by Green Rusts with Platinum

Figure 9(a) is the degradation kinetics of PCE by GR-SO_4 with or without of 0.5 mM Pt at pH 7.5. In the control sample, the initial concentration of the target compound decreased to 0.21 mM (89% of initial PCE concentration) at the first sampling point and it remained relatively constant. The initially reduced concentration of PCE was due to the sorption of PCE on the surface of septum lining and reactor wall and/or volatilization during the sample preparation process [67,68]. The concentration of PCE in GR-SO_4 suspension was not significantly decreased and remaining quite similar to that in the control sample during the reaction time (50 h). On the contrary, the reductive capacity for the degradation of PCE was remarkably increased in GR-SO_4(Pt) suspension. The target compound was fully degraded in 30 h and the formation of acetylene was conspicuously increased over the reaction time.

Arnold and Roberts proposed the reaction pathway for the reductive dechlorination of PCE that PCE can be initially transformed to TCE or dichloroacetylene and further degraded to acetylene, ethylene, and ethane [69]. Several transformation products such as TCE, acetylene, ethylene, and ethane were observed in the previous study on the PCE degradation with GR and GR with a transition metal [33,67]. However, no detectable amount of transformation products except acetylene were observed in this study and the total carbon mass balance of target and transformation products was approximately 98% at the last sampling point. In the Figure 9(a), the gray square represents acetylene concentration and the open circle indicates the sum of PCE and acetylene concentrations. Similar results were also reported on the rapid transformation of PCE to non chlorinated compounds, especially in the system containing reductants with a noble metal like Pd [70,71]. The results for the degradation of PCE by GR-SO_4(Pt) and formation of acetylene are shown in Figure 9(a) with solid lines. The degradation of PCE was fit well by the first-order kinetic model and the estimated rate constant was $0.0993 \pm 0.0058\ h^{-1}$, approximately two orders of magnitude higher than that of GR-SO_4 without Pt (Chart 6). The kinetic rate constant for the formation of acetylene was $0.0767 \pm 0.021\ h^{-1}$. Therefore, the Pt added GR in this study enhanced the reductive dechlorination of chlorinated aliphatic hydrocarbons similar to other transition metals (Ag, Au, and Cu) studied in the previous research [31-33].

The distribution of transformation products were not determined by the types of GRs.

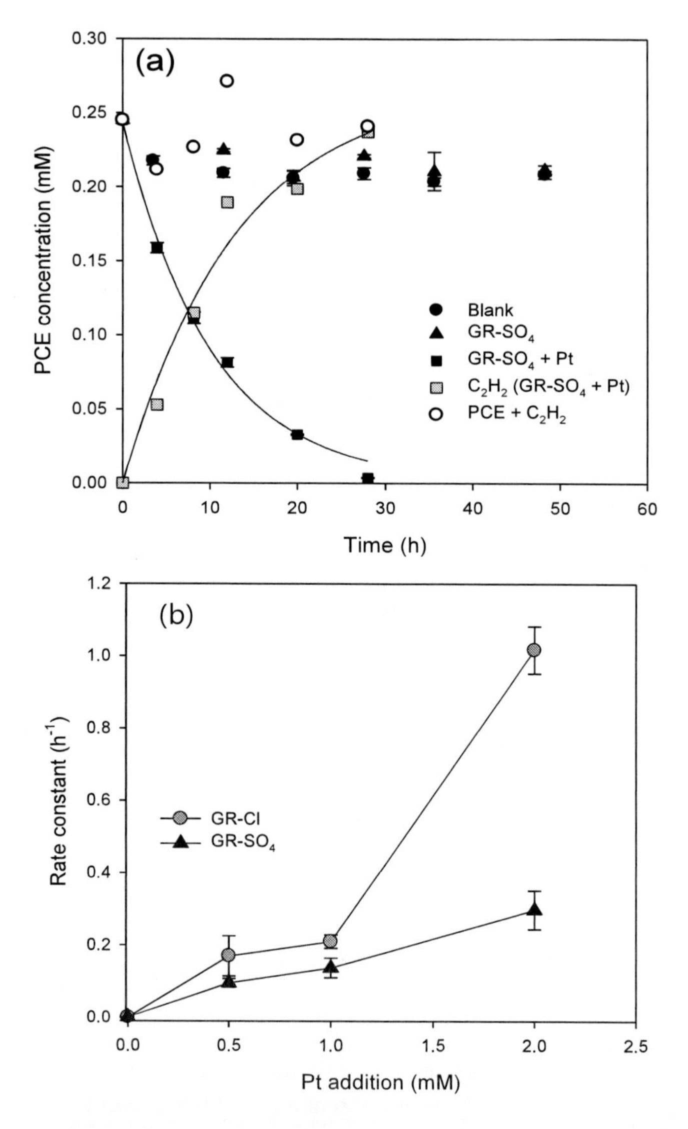

Figure 9. (a) Results of kinetic experiments on PCE degradation by $GR-SO_4$ in the absence and presence of 0.5 mM Pt at pH 7.5, (b) Dependence of degradation rates of GR-Cl(Pt) and $GR-SO_4$(Pt) on Pt concentration. The error bars for the rate constants in Fig. 1(c) represent the 95 % confidence intervals.

Only acetylene was formed as a transformation product in all experimental runs by GRs(Pt). At the last sampling point, the carbon mass balance (PCE + acetylene) varied in the range of 81% to 95%. The similar result was reported in the previous studies which have also found no significant dependence of the product distribution on the GR types. After reductive dechlorination of CT by GRs, 45.0% of CF and 4.0% of CH_4 transformed in GR-SO_4 suspension and 65.5% of CF and 8.7% of CH_4 transformed in GR-Cl suspension, respectively. However, no CF and 25% of CH_4 were measured in the GRs suspension modified with Cu [31,33]. From the experimental results in our research and the previous studies, therefore, we can propose that the distribution of chlorinated and non-chlorinated transformation products of PCE may depend on the presence of the transition metals in GR suspensions rather than the type of GRs.

Figure 9(b) shows the effect of Pt concentration on the reductive dechlorination kinetics of PCE. GR-Cl and GR-SO_4 were used to investigate the effect of Pt concentration with 0.5, 1, and 2 mM. The reductive dechlorination rates of PCE in both GR-Cl and GR-SO_4 suspensions were increased with respect to the Pt concentration. The slope of the estimated rate constants for the reductive dechlorination of the target compound by GR-SO_4(Pt) was 0.3 $h^{-1}mM^{-1}$ and it increased linearly. Compared to GR-SO_4(Pt), GR-Cl(Pt) indicated the substantial increase of rate constants with the increase of Pt concentration. The estimated rate constant of GR-Cl(Pt) increased from 0.21 h^{-1} to 1.02 h^{-1} when Pt concentration increased from 1 mM to 2 mM, which was 2.5 times greater than the increase in GR-SO_4. We suggest that the accelerated increase possibly because of its higher Fe(II) content in which more Pt bound on the surfaces of GR-Cl can play a reactive role in the reductive dechlorination of PCE. It might be helpful to understand the role of Pt in the GR suspensions that the observed effect of Pt concentration during the reductive dechlorination of PCE. Since tetra-valent Pt was added to the GRs suspensions in this experiment, it could not serve as a reductant for the reductive dechlorination of PCE. Rather, based on the effect of Pt concentration on the reductive dechlorination rate, Pt could play a role as a catalyst in assisting the electron transfer from GR surfaces to the target chlorinated compound [29,31]. In systems containing iron(II)-bearing soil minerals and zero-valent metals, it has been well recognized in the existing research that the catalytic role of transition metals in the reductive dechlorination of chlorinated compounds. For example, Ag(I), Au(III), and Cu(II) added into GR suspensions and Pd(II), Pt(IV), and Ni(II) into ZVI suspensions illustrated their catalytic roles in the strong reductive capacities for enhancing reductive dechlorination kinetics [29-31].

It has been speculated that the transition metals added into solid reductant suspensions are 1) initially attached on the surfaces of the solid reductants, 2) reduced to their zero-valent forms, and 3) degrade target chlorinated compounds adsorbed on the surfaces of the reductants, facilitating the electron transfer from the reductants to the target compounds [31,72].

The redox state of Pt on the surfaces of GR-Cl was verified using XPS analysis after the addition of Pt(IV) into the GR-Cl suspension. The concentration of Pt(IV) added into suspension was 10 mM because the XPS analysis could not exactly detect low levels of Pt concentration (i.e., $\leq$ 2 mM). The main chemical components of GR-Cl(Pt) (Fe, Cl, O, and Pt) were identified with the full scan spectra as shown in Figure 10. Figure 10(b)-(d) demonstrates the narrow scan of three elements (Pt (4f), Fe (2p3/2), and O (1s)) with their binding energies. In Figure 10(b), the two peaks at 71.27 and 74.39 eV in Pt (4f) spectrum show Pt(4f7/2) and Pt(4f5/2), respectively, which are consistent with those of metallic zero valent Pt (Pt(0)) [73]. The Pt(4f5/2) peak could not be resolved from the Al 2P peak of the Al foil used for sample drying. The reduced Pt(IV) to metallic Pt(0) clearly supports the hypothesis that the improved reductive dechlorination rates may be due to the catalytic role of Pt(0). The narrow region spectra for Fe 2p3/2 is showed in Figure 10(c) and it composed of three identical peaks at 709.5 (area percentage = 24.1%), 711.0 (37.8%), and 713.0 eV (38.1%), respectively. It has been reported that the binding energy for Fe(II)-O and Fe(III)-O to be in the range of 709 to 709.5 eV and 711 to 714 eV, respectively [61,62]. Therefore, we can conclude that iron bound on the GR surface with Pt was consisted of approximately 24 % of Fe(II) and 76 % of Fe(III). The substantially lower ratio between Fe(II) to Fe(III) of GR-Cl(Pt) than that of GR-Cl indicate that Pt(IV) on the GR-Cl surfaces was reduced to Pt(0) via coupled surface Fe(II) oxidation by electron-transferring; during the sample preparation, atmospheric oxygen may also contribute to the oxidation of GR-Cl(Pt) resulting in the low Fe(II) content on the GR surfaces. Figure 10(d) shows the result of O 1s spectrum that can be resolved into two peaks at 529.68 and 531.42 eV representing O^{2-} and OH^-, respectively [61,62]. The presence of O^{2-} species indicate that GR may be transformed to other iron bearing minerals, possibly iron oxide, during the reaction with Pt, based on that GR consists of Fe(II)-Fe(III) hydroxide layer and anions in interlayer.

It is investigated by SEM that the micro-morphology of zero-valent Pt on the surface of GR-Cl(Pt) and GR-SO_4(Pt).

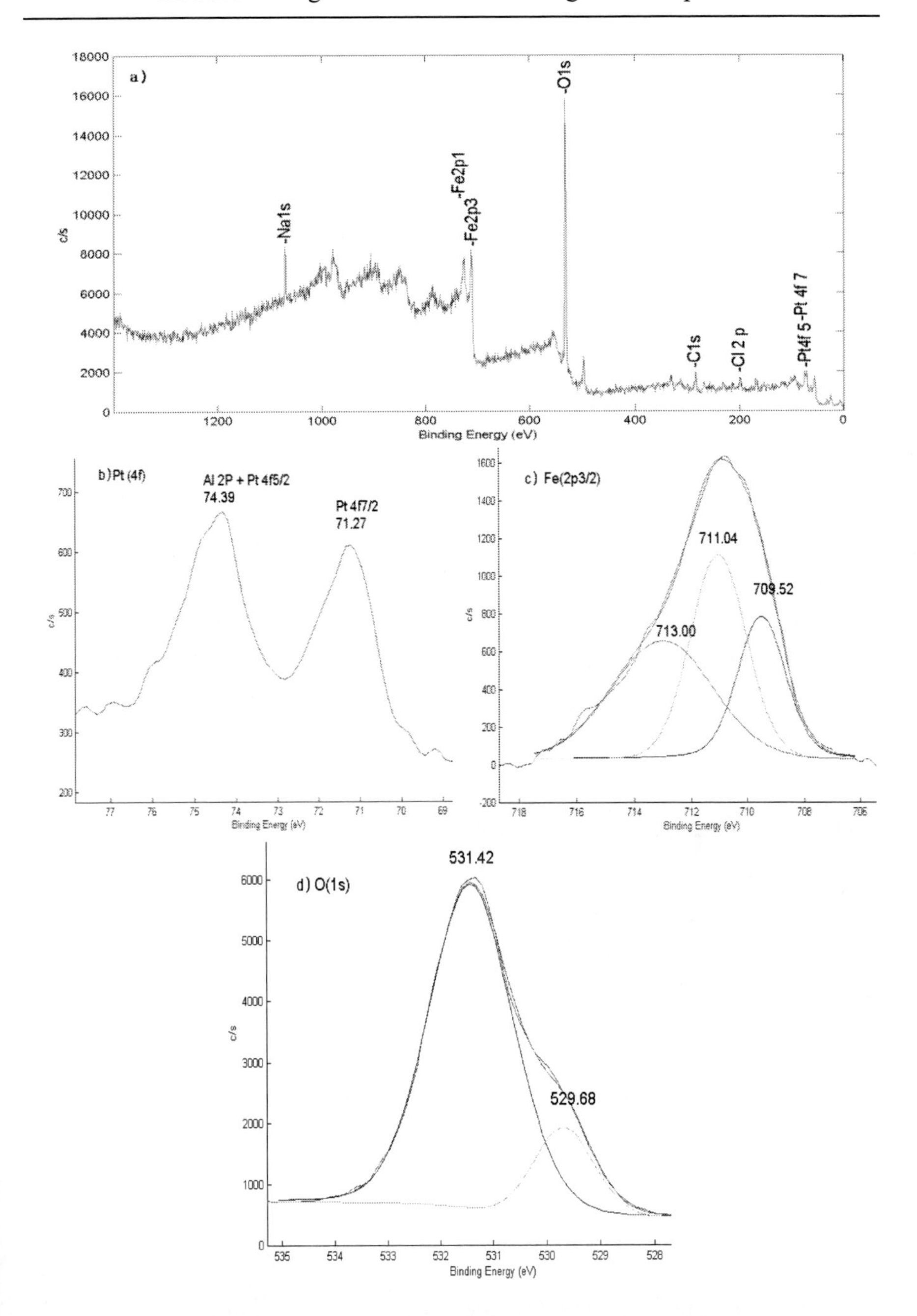

Figure 10. Resutls of surfac analyses (XPS and XRD) of GR-Cl(Pt). XPS analysis of GR-Cl(Pt): (a) full scan spectrum, (b) narrow region scans of Pt(4f7/2), (c) Fe(2p3/2), and (d) O(1s), respectively.

Figure 11(a) and (b) show that hundreds of nano-sized particles (see regions within the circles) have dark gray hexagonal shapes associated onto the surfaces of GR-Cl. EDS analysis was conducted on the surfaces and it shows main chemical components are O, Cl, Fe, and Pt as shown in Figure 11(a) and (b). From these surface analyses results by XPS and SEM with EDS, therefore, we can conclude that the nano-sized particles on the GR surfaces are zero valent Pt. O'Loughlin [74] group reported that nano-sized zero valent transition metals can be formed on the surface of reactive solids. Figure 10(c), at 10 times lower magnification than Figure 10(a), shows the bright Pt(0) particles in the circles were aggregated on the GR surfaces.

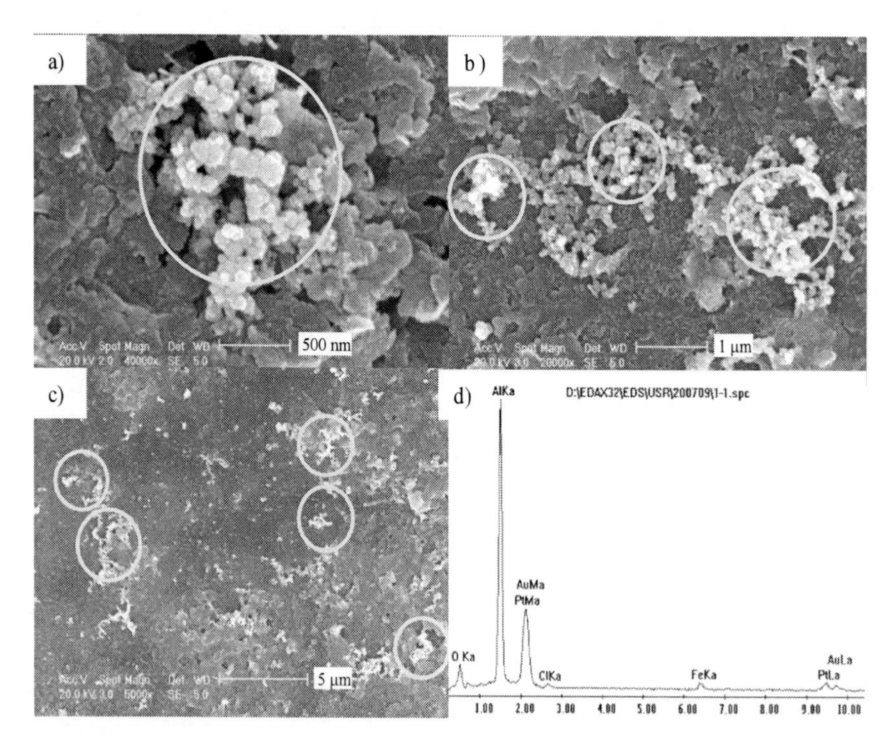

Figure 11. (a,b, and c) SEM image and (d) EDS spectra of GR-Cl(Pt).

IV. Environmental Significance

COCs classified as one of the toxic VOCs group are widespread groundwater and soil contaminants in our living site. Their toxicity to humans and animals and persistence in natural environments have been known to everywhere throughout the literature. In order to control and remove those hazardous chlorinated organic chemicals in the natural and engineered system, many advanced technologies have been developed such as advanced oxidation processes, nano-sized reactive materials and electronic-chemical remediation technologies. Recently, researchers has been shed new light on natural attenuation as increase of more economic and environmental friendly concern. Especially, iron bearing soil minerals and phyllosilicates have been attracted for the reductive degradation of COCs at natural attenuation process due to its reduction capacity to contaminants. Fe(II) contained and/or added in iron bearing soil mineral and phyllosilicate suspensions can be a source of electron donor to degrade COCs. The experimental results as shown in this chapter have pointed out the significant application of iron bearing soil minerals and phyllosilicates on the reductive dechlorination of COCs. The reductive capacity of iron bearing natural materials for COCs examined in this chapter can provide important background knowledge to understand degradation mechanism of COCs at the real site fully contained by iron bearing soil minerals and phyllosilicates and to estimate removal efficiency and operating time during the natural attenuation or other related remedial technologies. In addition, the enhanced degradation rate of COCs by addition of transition metals on the reductive dechlorination of COCs by iron bearing soil minerals indicates that catalytic effect of transition metals on iron bearing soil minerals can be used to accelerate removal efficiency and rate at natural attenuation process, which showed low removal efficiency and slow degradation rate. Although the results obtained from this research using batch test could not provide all environmental phenomena during the reductive dechlorination of COCs by iron bearing soil minerals and phyllosilicates, it would be helpful to provide an approximate estimation to predict contaminants transformation in natural attenuation and in-situ redox manipulation.

Acknowledgments

The research in this chapter was partially funded by the State of Texas as part of the program of the Texas Hazardous Waste Research Center, Korean Ministry of Science and Technology (R01-2006-000-10727-0), Korea Research Foundation (KRF-2007-313-D00439), Korean Ministry of Land, and Transport and Maritime Affairs (06-CC-A01: High-tech Urban Development Program).

REFERENCES

[1] Hutzinger, O.; Veerkamp, W. In *Microbial Degradation of Xenobiotic and Recalcitrant Compounds*; Leisinger, T., Ed.; Academic Press: London, 1981.

[2] Hileman, B. Concerns broaden over chlorine and chlorinated hydrocarbons. *Chem. Engr. News* 1993, *19*, 11-20.

[3] Ukrainczyk, L.; Chibwe, M.; Pinnavaia, T. J.; Boyd, S. A. Reductive Dechlorination of Carbon Tetrachloride In Water Catalyzed by Mineral-Supported Biomimetic Cobalt Macrocycles. *Environ. Sci. Technol.* 1995, *29*, 439-445.

[4] Leisinger, T. In: *Biotechnology*; Rehm, H. J., Reed, G., Eds.; VCH Verlagsgesellschaft: Weinheim, 1986; Vol. 8, pp 475 - 513.

[5] Fathepure, B. Z.; Boyd, S. A. Dependence of tetrachloroethylene dechlorination of methanogenic substrate consumption by Methan osarcina sp. strain DCM. *App. Environ. Microbiol.* 1988, *54*, 2976-2980.

[6] McCarty, P. L. Biotic and Abiotic Transformations of Chlorinated Solvents in Ground Water. Ward, C. H., Ed.; In *Symposium on Natural Attenuation of Chlorinated Organics in Ground Water*; EPA: Dallas, TX, 1996; pp 5-9.

[7] Holliger, C.; Schraa, G.; Stams, A. J. M.; Zehnder, A. J. B. A highly purified enrichment culture couples the reductive dechlorination of tetrachloroethene to growth. *App. Environ. Microbiol.* 1993, *59*, 2991-2997.

[8] Schwarzenbach, R. P.; Giger, W.; Schaffner, C.; Wanner, O. Groundwater contamination by volatile halogenated alkanes: abiotic formation of volatile sulfur compounds under anaerobic conditions. *Environ. Sci. Technol.* 1985, *19*, 322-327.

[9] Bagley, D. M.; Gossett, J. M. Tetrachloroethene transformation to trichloroethene and cis-1,2-dichloroethene by sulfate-reducing enrichment cultures. *App. Environ. Microbiol.* 1990, *56*, 2511-2516.

[10] Freedman, D. L.; Gossett, J. M. Biological reductive dechlorination of tetrachloroethylene and trichloroethylene to ethylene under methanogenic conditions. *App. Environ. Microbiol.* 1989, *55*, 2144-2151.

[11] Fathepure, B. Z.; Nengu, J. P.; Boyd, S. A. Anaerobic bacteria that dechlorinate perchloroethene. *App. Environ. Microbiol.* 1987, *53*, 2671-2674.

[12] Gossett, J. M. Microbiological Aspects Relevant to Natural Attenuation of Chlorinated Ethenes. Ward, C. H., Ed.; In *Symposium on Natural*

Attenuation of Chlorinated Organics in Ground Water; EPA: Dallas, TX, 1996; pp 10-13.

[13] Holliger, C.; Schumacher, W. Reductive dehalogenation as a respiratory process. *Anton. Van Lee.* 1994, *66*, 239-246.

[14] Butler, E. C. Ph.D. Dissertation, University of Michigan: Ann Arbor, MI, 1998.

[15] Weaver, J. W.; Wilson, J. T.; Kampbell, D. H. Case Study of Natural Attenuation of Trichloroethene at St. Joseph, Michigan. Ward, C. H., Ed.; In *Symposium on Natural Attenuation of Chlorinated Organics in Ground Water*; EPA: Dallas, TX, 1996; pp 69-73.

[16] Sokol R. C.; Kwon O.; Bethoney C. M.; Rhee G. Reductive dechlorination of polychlorinated biphenyls in St. Lawrence river sediments and variations in dechlorination characteristics. *Environ. Sci. Technol.* 1994, *28*, 2054-2056.

[17] Kriegman-King, M. R.; Reinhard, M. *Abiotic Transformation of Carbon Tetrachloride at Mineral Surfaces*; EPA/600/SR-49/018; EPA: Ada, OK, 1994.

[18] Kriegman-King, M. R. Transformation of Carbon Tetrachloride by Pyrite in Aqueous Solution. *Environ. Sci. Technol.* 1994, *28*, 692-700.

[19] Sivavec, T. M.; Horney, D. P. Reduction of Chlorinated Solvents by Fe(II) Minerals. 213th ACS National Meeting; ACS, 1997, pp 115-117.

[20] Butler, E. C.; Hayes, K. F. Effects of Solution Composition and pH on the Reductive Dechlorination of Hexachloroethane by Iron Sulfide. *Environ. Sci. Technol.* 1998, *32*, 1276-1284.

[21] Butler, E. C.; Hayes, K. F. Kinetics of the Transformation of Trichloroethylene and Tetrachloroethylene by Iron Sulfide. *Environ. Sci. Technol.* 1999, *33*, 2021-2027.

[22] Lee, W.; Batchelor, B. Abiotic Reductive Dechlorination of Chlorinated Ethylenes by Iron Bearing Soil Minerals and Potential Interactions with Biotic Processes. In: *Chemical-Biological Interactions in Contaminant Fate*; Tratnyek, P. G., Adriaens, P., Roden, E. E., Eds.; 220th ACS National Meeting; ACS: Washington, DC, 2000, pp 338-340.

[23] Erbs, M.; Hansen, H. C. B.; Olsen, C. E. Reductive Dechlorination of Carbon Tetrachloride Using Iron(II) Iron(III) Hydroxide Sulfate (Green Rust). *Environ. Sci. Technol.* 1999, *33*, 307-311.

[24] O'Loughlin, E. J.; Burris, D. R. Reductive Transformation of Halogenated Hydrocarbons by Green Rust. 220th ACS National Meeting; ACS: Washington, DC, 2000, pp 635-636.

[25] Kriegman-King, M. R.; Reinhard, M. Transformation of carbon tetrachloride in the presence of sulfide, biotite, and vermiculite. *Environ. Sci. Technol.* 1992, 26, 2198-2206.

[26] Fennelly, J. P.; Roberts, A.L. Reaction of 1,1,1-trichloroethane with zero-valent metals and bimetallic reductants. *Environ. Sci. Technol.* 1998, *32*, 1980–1988.

[27] Wan, C.; Chen, Y.H.; Wei, R. Dechlorination of chloromethanes on iron and palladium–iron bimetallic surface in aqueous systems. *Environ. Toxicol. Chem.* 1999, *18*, 1091–1096.

[28] Kim, Y; Caraway, E.R. Dechlorination of pentachlorophenol by zero valent iron and modified zero valent irons, *Environ. Sci. Technol.* 2000, *34*, 2014–2017.

[29] Zhang, W.; Wang, C.; Lien, H. Treatment of chlorinated organic contaminants with nanoscale bimetallic particles, *Catal. Today*. 1998, *40*, 387–395.

[30] Cheng, S.-F.; We, S.-C. The enhancement methods for the degradation of TCE by zero-valent metals, *Chemosphere*. 2000, *41*, 1263–1270.

[31] O'Loughlin, E. J.; Kemner, K. M.; Burris, D. R. Effects of Ag(I), Au(III), and Cu(II) on the reductive dechlorination of carbon tetrachloride by green rust. *Environ. Sci. Technol.* 2003, *37*, 2905-2912.

[32] O'Loughlin, E. J.; Burris, D. R. Reduction of halogenated ethanes by green rust. *Environ. Toxicol. Chem.* 2004, *23*, 41-48.

[33] Maithreepala, R. A.; Doong, R.-A. Enhanced dechlorination of chlorinated methanes and ethenes by chloride green rust in the presence of copper(II). *Environ. Sci. Technol.* 2005, *39*, 4082-4090.

[34] Taylor, R. M.; Maher, B. A.; Self, P. G. Magnetite in soils; I, The synthesis of single-domain and superparamagnetic magnetite. *Clay Miner.* 1987, *22*, 411-422.

[35] Kriegman-King, M. R. Ph.D. Dissertation, Stanford University: Stanford, CA, 1993.

[36] Drissi, S. H.; Refait, P.; Abdelmoula, M.; Genin, J. M. R. The preparation and thermodynamic properties of Fe(II)-Fe(III) hydroxide-carbonate (green rust 1) ; pourbaix diagram of Iron in carbonate-containing aqueous media. *Corros. Sci.* 1995, *37*, 2025-2041.

[37] Olowe, A. A.; Genin, J. M. G. The mechanism of oxidation of ferrous hydroxide in sulphated aqueous media: Importance of the initial ratio of the reactants. *Corros. Sci.* 1991, *32*, 965-984.

[38] Refait, P. H.; Drissi, S. H.; Pytkiewicz, J.; Genin, J. M. R. The Anionic species competition in iron aqueous corrosion: role of various green rust compounds. *Corros. Sci.* 1997, *39*, 1699-1710.
[39] Gossett, J. M. Measurement of Henry's law constants for C1 and C2 chlorinated hydrocarbons. *Environ. Sci. Technol.* 1987, *21*, 202-208.
[40] Lee, W.; Batchelor, B. Reductive capacity of natural reductants, *Environ. Sci.Technol.* 2003, *37*, 535–541.
[41] Arnold, W. A.; Roberts, A. L. Pathways of Chlorinated Ethylene and Chlorinated Acetylene Reaction with Zn(0). *Environ. Sci. Technol.* 1998, *32*, 3017-3025.
[42] Mackay, D.; Shiu, W. Y. A critical review of Henry's law constants for chemicals of environmental interest. *J. Phys. Chem. Ref. Data* 1981, *10*, 1175-1199.
[43] Klausen, J.; Tröber, S. P.; Haderlein, S. B.; Schwarzenbach, R. P. Reduction of Substituted Nitrobenzenes by Fe(II) in Aqueous Mineral Suspensions. *Environ. Sci. Technol.* 1995, *29*, 2396-2404.
[44] Roberts, A. L.; Totten, L. A.; Arnold, W. A.; Burris, D. R.; Campbell, T. J. Reductive Elimination of Chlorinated Ethylenes by Zero-Valent Metals. *Environ. Sci. Technol.* 1996, *30*, 2654-2659.
[45] Bradley, P. M.; Chapelle, F. H. Kinetics of DCE and VC Mineralization under Methanogenic and Fe(III)-Reducing Conditions. *Environ. Sci. Technol.* 1997, *31*, 2692-2696.
[46] Sivavec, T. M.; Park, C. Inventors; General Electric Company, Assignee; Composition and method for ground water remediation. USA patent 5750036. May, 1998.
[47] McCormick, M. L.; Kim, H. S.; Adriaens, P. Transformation of tetrachloromethane in a defined iron reducing culture: relative contributions of cell and mineral mediated reactions. in: *Specialty Chemicals in the Environment*; Stone, A. T., Ed.; 219th ACS National Meeting; ACS: San Francisco, CA, 2000, pp 138-141.
[48] Lee, W. Ph.D. Dissertation, Texas A&M University: College Station, TX, 2001.
[49] Asami, K.; Hashimoto, K. The X-ray photo-electron spectra of several oxides of iron and chromium. *Corrosion Sci.* 1977, *17*, 559-570.
[50] McCafferty, E.; Bernett, M.K.; Murday, J.S. An XPS study of passive film formation on iron in chromate solutions. *Corro. Sci.* 1988, *28*, 559-576.

[51] Lowry, G.V.; Reinhard, M. Hydrodehalogenation of 1- to 3-carbon halogenated organic compounds in water using a palladium catalyst and hydrogen gas. *Environ. Sci. Technol.* 1999, *33*, 1905–1910.

[52] Doong, R.-A.; Chen, K.-T.; Tsai, H.-C. Reductive dechlorination of carbon tetrachloride and tetrachloroethylene by zerovalent silicon-iron reductants. *Environ. Sci. Technol.* 2003, *37*, 2575–2581.

[53] Danielsen, K.M.; Hayes, K.F. pH dependence of carbon tetrachloride reductive dechlorination by magnetite. *Environ. Sci. Technol.* 2004, *38*, 4745–4752.

[54] Zwank, L.; Elsner, M.; Aeberhard, A.; Schwarzenbach, R.P. Carbon isotope fractionation in the reductive dehalogenation of carbon tetrachloride at iron (hydr)oxide and iron sulfide minerals. *Environ. Sci. Technol.* 2005, *39*, 5634–5641.

[55] Tamara, M.L.; Butler, E.C. Effects of iron purity and groundwater characteristics on rates and products in the degradation of carbon tetrachloride by iron metal. *Environ. Sci. Technol.* 2004, *38*, 1866–1876.

[56] Gander, J.W.; Parkin, G.F.; Scherer, M.M. Kinetics of 1,1,1-trichloroethane transformation by iron sulfide and 1 methanogenic consortium. *Environ. Sci. Technol.* 2002, *36*, 4540–4546.

[57] Butler, E.C.; Hayes, K.F. Kinetics of the transformation of halogenated aliphatic compounds by iron sulfide. *Environ. Sci. Technol.* 2000, 34, 422–429.

[58] Jeong, H.Y.; Hayes, K.F. Impact of transition metals on reductive dechlorination rate of hexachloroethane by mackinawite. *Environ. Sci. Technol.* 2003, 37, 4650–4655.

[59] Lin, C.J.; Lo, S.L.; Liou,Y.H. Dechlorination of trichloroethylene in aqueous solution by noble metal-modified iron. *J. Hazard. Mater.* 2004, *B116*, 219–228.

[60] Wilkin, R.T.; Bames, H.L. Formation processes of framboidal pyrite. *Geochim. Cosmochim. Acta* 1997, *61*, 323–339.

[61] Thomas, J.E.; Jones, C.F.; Skinner, W.M.; Smart, R.S.C. The role of surface sulfur species in the inhibition of pyrrhotite dissolution in acid conditions. *Geochim. Cosmochim. Acta* 1998, *62*, 1555–1565.

[62] Mullet, M.; Boursiquot, S.; Abdelmoula, M.; Genin, J.-M.; Ehrhardt, J.-J. Surface chemistry and structural properties of mackinawite prepared by reaction of sulfide ions with metallic iron. *Geochim. Cosmochim. Acta* 2002, *66*, 829–836.

[63] Herbert, R.B.; Benner, S.G.; Pratt, A.R.; Blowes, D.W. Surface chemistry and morphology poorly crystalline iron sulfides precipitated in

media containing sulfate-reducing bacteria. *Chem. Geol.* 1998, *144*, 87–97.

[64] Juskenas, R.; Valsiunasa, I.; Pakstasa, V.; Selskisa, A.; Jasulaitienea, V.; Karpavicienea, V.; Kapociusa, V. XRD, XPS and AFM studies of the unknown phase formed on the surface during electrodeposition of Ni–W alloy. *Appl. Surf. Sci.* 2006, *253*, 1435–1442.

[65] Vaughan, D.J.; Craig, J.R. Mineral Chemistry of Metal Sulfides, Cambridge University Press, New York, 1978.

[66] Meng, Z.; Peng, Y.; Yu, W.; Qian, Y. Solvothermal synthesis and phase control of nickel sulfides with different morphologies. *Mater. Chem. Phys.* 2002, *74*, 230–233.

[67] Lee, W.; Batchelor, B. Abiotic reductive dechlorination of chlorinated ethylenes by iron-bearing soil minerals. 2. Green rust. *Environ. Sci. Technol.* 2002, *36*, 5348-5354.

[68] Lee, W.; Batchelor, B. Abiotic reductive dechlorination of chlorinated ethylene by iron-bearing soil minerals. 1. pyrite and magnetite. *Environ. Sci. Technol.* 2002, *36*, 5147-5154.

[69] Arnold, W. A.; Roberts, A. L. Pathways and kinetics of chlorinated ethylene and chlorinated acetylene reaction with Fe(0) particles. *Environ. Sci. Technol.* 2000, *34*, 1794-1805.

[70] Kim, Y. H.; Carraway, E. R. Dechlorination of chlorinated ethenes and acetylenes by palladized iron. *Environ. Technol.* 2003, *24*, 809-819.

[71] Lien, H.-L.; Zhang, W.-X. Nanoscale iron particles for complete reduction of chlorinated ethenes. *Colloids Surf. A: Physicochem. Eng. Aspects* 2001, *191*, 97-105.

[72] Cheng, I. F.; Fernando, Q.; Korte, N. Electrochemical dechlorination of 4-chlorophenol to phenol. *Environ. Sci. Technol.* 1997, *31*, 1074-1078.

[73] M'Boungou, J. S.; Hilaire, L.; Maire, G.; Garin, F. Role of tungsten in supported Pt-W reforming catalysts Part II. Influence of the tungsten loading and metal-support interactions effect. *Catal. Lett.* 1991, *10*, 401-411.

[74] O'loughlin, E. J.; Kelly, S. D.; Kemner, K. M.; Csencsits, R.; Cook, R. E. Reduction of Ag^{I}, Au^{III}, Cu^{II}, and Hg^{II} by Fe^{II}/Fe^{III} hydroxysulfate green rust. *Chemosphere* 2003, *53*, 437-446.

In: Volatile Organic Compounds ISBN 978-1-61324-156-1
Editors: J. C. Hanks et al. pp. 47-88

Chapter 2

PHOTOCATALYTIC DEGRADATION OF VOC GASES CONSIDERING REACTOR DESIGN FOR THE TREATMENT OF DECOMPOSITION INTERMEDIATES USING A SHORT WAVELENGTH UV LIGHT AND WATER DROPLETS

Kazuhiko Sekiguchi and Kyung Hwan Kim
Division of Environmental Science and Infrastructure Engineering,
Graduate School of Science and Engineering,
Saitama University, Sakura, Saitama, Japan

ABSTRACT

Volatile organic compounds (VOCs) are of environmental concern because of their adverse effects on human health. Moreover, VOCs can adversely affect certain manufacturing processes; for example, in semiconductor manufacturing, the wafer surface is damaged by VOCs. Although chemical filtration of classical removal techniques are effective in removing a variety of hazardous VOCs from air, these methods are expensive, and the materials used for adsorption and filtration have short and unpredictable life spans. Accordingly, there is much interest in developing new air purification systems for removing VOCs from air. In this Chapter, we introduce the photodegradation of toluene and benzene with TiO_2 and $UV_{254+185nm}$ irradiation from an ozone-producing UV lamp as a first step to show VOC degradation in gas phase using short-wavelength UV irradiation with TiO_2 catalyst with different UV sources. The results show that VOCs were decomposed and mineralized efficiently owing to the synergetic effect of photochemical oxidation in gas phase and photocatalytic oxidation on the TiO_2 surface. The conversion levels obtained with $UV_{254+185nm}$ photoirradiated TiO_2 were much higher than those obtained with conventional UV sources (UV_{365nm} and UV_{254nm}), which suffer from both catalyst deactivation and the generation of harmful intermediates. The products from the photo degradation of VOCs with the $UV_{254+185nm}$ photoirradiated TiO_2 were mainly mineralized CO_2 and CO, but some water-soluble organic intermediates were also formed under more severe reaction conditions. The water-soluble aldehydes and carboxylic intermediates disappeared from the effluent gas stream and were detected in the water impingers. These findings suggest that the intermediates can be washed out by conventional gas washing technique, such as wet scrubber or air washer. As a second step, we introduced removal of water-insoluble gaseous pollutants (NOx and toluene) with $UV_{254+185nm}$ irradiation in humid air to provide a useful process for effectively removing gaseous pollutants from the air and the treatment of intermediates by trapping the water-soluble intermediates into water droplets by the air washer. The results show that the OH radicals and ozone produced by $UV_{254+185nm}$ irradiation effectively degraded NOx and toluene to HNO_3 and CO_2, respectively. The organic intermediates formed during toluene degradation were highly water soluble and could therefore be effectively removed along with the HNO_3, by the air washer. However, using the air washer as a means for effective removal of gaseous pollutants and their intermediates has disadvantages for example, 1) the reaction can be completed with 2-step process and 2) the size of air washer is relatively large. For these reasons, finally we studied and described VOCs degradation using an ultrasonic mist generated from TiO_2 suspension under UV_{365nm}, UV_{254nm}, and $UV_{254+185nm}$ irradiations. With this technology, gaseous pollutants and the intermediates could be degraded and captured by water droplet and decomposed further in a liquid phase with 1-step process by generating mist droplets from ultrasonic atomization of TiO_2 suspension. Organic gas species, UV wavelength, the diameter of ultrasonic mist containing photocatalyst particles (UMP), and the inhibition of the formation of secondary particles from intermediates were also investigated. In this method, VOCs were decomposed on the surface or inside of UMP and the degradation intermediates from hydrophobic gas could also be captured by water droplet and decomposed further in a liquid phase.

1. INTRODUCTION

Volatile organic compounds (VOCs) are of environmental concern because of their adverse effects on human health. Moreover, VOCs can adversely affect certain manufacturing processes; for example, in semiconductor manufacturing, the wafer surface is damaged by VOCs. Although chemical filtration of classical removal techniques are effective in removing a variety of hazardous VOCs from air, these methods are expensive, and the materials used for adsorption and filtration have short and unpredictable life spans. Accordingly, there is much interest in developing new air purification systems for removing VOCs from air.

In the past decade, heterogeneous photocatalytic oxidation (PCO) has been extensively studied as a promising method for removing toxic organic compounds from air. PCO is cost-effective and can be carried out at room temperature and atmospheric pressure, with good catalyst stability. However, application of PCO in industry has been problematic. Photocatalytic reaction depends on the chemical properties of the pollutants; for example, the conversion of aromatic compounds, such as toluene, is more difficult than that of chlorinated hydrocarbons or alcohols [1]. Some researchers have reported that the reaction efficiency of BTEX (benzene, toluene, ethylbenzene, and xylenes) depends on the amounts of BTEX adsorbed on the titanium dioxide (TiO_2) catalyst [2,3]. Deactivation of the TiO_2 catalyst is also a major disadvantage in the industrial application of PCO. During the PCO of aromatic compounds, less-reactive intermediates are directly responsible for deactivation of the catalyst. These intermediates are strongly adsorbed on the surface of the TiO_2 catalyst and deteriorate photoactivity by blocking reaction sites [4-8]. Generation of undesirable intermediates can also strongly inhibit the application of PCO. Moreover, since some intermediates are more toxic than their parent compounds, these intermediates must be completely removed from the effluent gas stream [9]. Thus, many studies have been devoted to eliminate these disadvantages of photocatalytic degradation. However, most of these studies have been mainly focused on catalyst modification (e.g., synthesis of a nanosized TiO_2 photocatalyst, noble metal deposition or ion doping, or synthesis of a visible-light photocatalyst) [10-13].

In general, black light blue lamp with wavelengths in the 320–400 nm region (UV-A) is used as light source to provide energy in TiO_2-catalyzed photoreactions. Wavelengths below 315 nm are rarely used [14]. Particularly, no study has been reported for removing VOCs in polluted air using short-wavelength $UV_{254+185nm}$ with TiO_2 catalyst, except for some applications for treatment of wastewater [15, 16]. In fact, a light source with a wavelength in the 320–400 nm regions is sufficient to promote electrons from the valence band to the conduction band of the TiO_2 particles. However, some studies have shown that the $UV_{254+185nm}$ irradiation could generate a great number of a reactive species like hydroxyls radical and ozone in the air [17-19]; hence it is expected that if the $UV_{254+185\ nm}$ irradiation can efficiently excite the TiO_2 catalyst, the combination will significantly decompose VOCs with aid of combined effect of photochemical oxidation in the gas phase and photocatalytic oxidation on the TiO_2 catalyst.

In the first study, we evaluated the application of short wavelength UV as light source of TiO_2 catalyst for photodegradation of gaseous VOCs to improve the disadvantages associated with classical photocatalytic degradation. We characterized the photochemical and photocatalytic oxidation by using the TiO_2 catalyst under $UV_{254+185nm}$ irradiation, and investigated the synergistic effect of the combination. The photodegradation of gaseous toluene by different UV (365, 254, and 254 + 185 nm) irradiations was compared. The results show that VOCs were decomposed and mineralized efficiently owing to the synergetic effect of photochemical oxidation in gas phase and photocatalytic oxidation on the TiO_2 surface. The conversion levels obtained under $UV_{254+185nm}$ irradiation with TiO_2 were much higher than those obtained with conventional UV sources (UV_{365nm} and UV_{254nm}), which suffer from both catalyst deactivation and the generation of harmful intermediates. The products from the photodegradation of VOCs under $UV_{254+185nm}$ irradiation with TiO_2 were mainly mineralized CO_2 and CO, but some water-soluble organic intermediates were also formed under more severe reaction conditions. The water-soluble aldehydes and carboxylic intermediates disappeared from the effluent gas stream and were detected in the water impingers. These findings suggest that the intermediates can be washed out by conventional gas washing technique, such as wet scrubber or air washer.

Therefore, we introduced removal of water-insoluble gaseous pollutants (NOx and toluene) with $UV_{254+185nm}$ irradiation in humid air to provide a useful process for effectively removing gaseous pollutants from the air and the treatment of intermediates by trapping the water-soluble intermediates into water droplets by the air washer as a second approach. The air washers, which are widely used to adjust indoor humidity in clean rooms, have recently received considerable attention for removing gaseous pollutants. The air washer can effectively remove water-soluble pollutants such as SO_2 and NH_3 by means of a fine mist sprayed through a nozzle, but water-insoluble pollutants are difficult to remove. However, converting water insoluble pollutants into water-soluble materials would increase the number of gaseous pollutants that could be removed with an air washer. This study describes an irradiation process that uses a short UV wavelength ($UV_{254+185nm}$) to convert water-insoluble pollutants to water-soluble materials or to harmless substances. When humid air is irradiated with $UV_{254+185nm}$, large quantities of ozone and hydroxyl radicals are produced in the air stream [19, 20]. In the previous studies, we have reported that $UV_{254+185nm}$ irradiation efficiently decomposes some aromatic organic compounds to CO_2 and/or converts them into more water-soluble products such as aldehydes or carboxylic acids [21, 22]. Thus in this second study, we chose water-insoluble NOx (NO and NO_2) and toluene as model pollutants to investigate the water-solubility of the decomposition products. A laboratory-scale air washer was used for post-treatment of the water-soluble decomposition products. We also examined the possibility of enhancing the photodegradation of the pollutants by carrying out the $UV_{254+185\ nm}$ irradiation in a reactor dipcoated with TiO_2. The results show that the OH radicals and ozone produced by $UV_{254+185nm}$ irradiation effectively degraded NOx and toluene to HNO_3 and CO_2, respectively. The organic intermediates formed during toluene degradation were highly water soluble and could therefore be effectively removed along with the HNO_3, by the air washer.

However, using the air washer as a means for effective removal of gaseous pollutants and their intermediates has disadvantages for example, 1) the reaction can be completed with 2-step process and 2) the size of air washer is relatively large. With these reasons, finally we studied and described VOCs degradation using an ultrasonic mist generated from TiO_2 suspension under UV_{365nm}, UV_{254nm}, and $UV_{254+185nm}$ irradiations. With this technology, gaseous pollutants and the intermediates could be degraded and captured by water droplet and decomposed further in a liquid phase with 1-step process by generating mist droplets from ultrasonic atomization of TiO_2 suspension. This technique is useful for producing fine water mist with particle diameters less than 10 μm and does not require large amounts of energy compared with heating and irradiation frequencies in the high ultrasonic range (2.4 MHz). In our previous study [22], TiO_2 photocatalyst particles were introduced into a fine water mist by ultrasonic atomization of a TiO_2 suspension. Using this method, we could photocatalytically decompose toluene gas under stable relative humidity conditions. This ultrasonic mist (UM) containing the photocatalyst particles was referred to as ''UMP''. Organic gas species, UV wavelength, the diameter of ultrasonic mist containing UMP, and the inhibition of the formation of secondary particles from intermediates were also investigated. In this method, VOCs were decomposed on the surface or inside of UMP and the degradation intermediates from hydrophobic gas could also be captured by water droplet and decomposed further in a liquid phase. The effectiveness of this proposed technique as an air purification method is introduced and discussed.

In this third study, we have investigated techniques for promoting the degradation of pollutants via photochemical reaction in gas phase and photocatalytic reaction on mist surface using several short-wavelength UV sources, as well as for treating the intermediate degradation products after the reaction using an air washer with sprayed water [23-25]. From these studies, it was found that the intermediate products of VOC degradation were water soluble and could be rapidly trapped in water mist. Furthermore, we have proposed a system using ultrasonic atomization to promote the simultaneous photocatalytic degradation and trapping of intermediate degradation products in water [22].

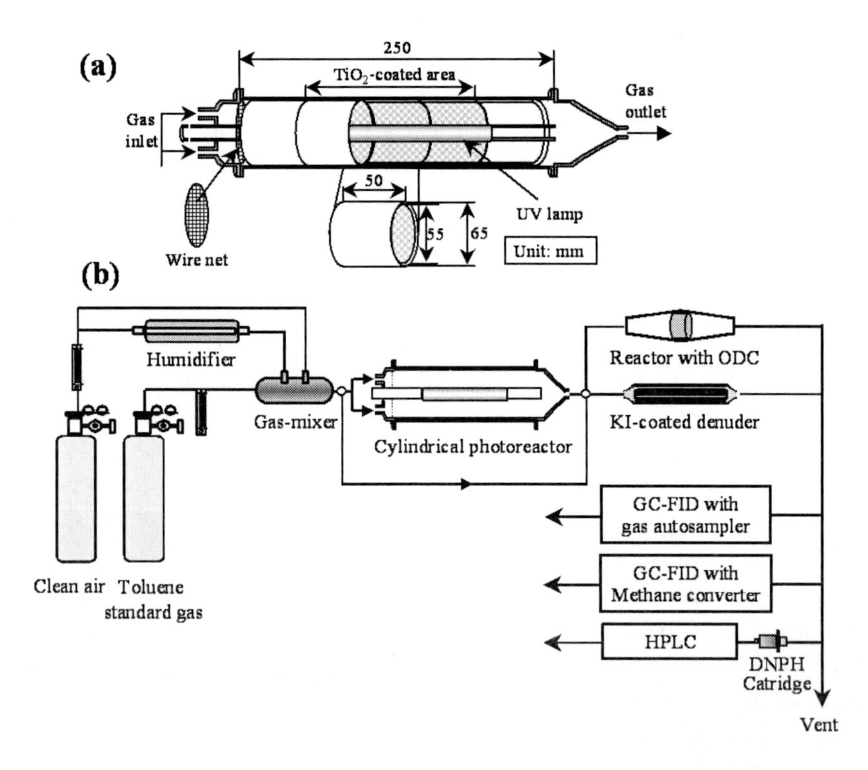

Figure 1. (a) The cylindrical photoreactor. (b) Schematic diagram of the experimental apparatus.

2. EXPERIMENTAL AND METHODS

2.1. Application of $UV_{254+185nm}$ Irradiation to the VOC Degradation using TiO_2 Photocatalyst

2.1.1. Phtoreactor and UV Sources

The photoreactor (V = 0.55 L) consisted of a Pyrex glass cylinder with five inner cylinders (Figure 1(a)). The three innermost cylinders were coated with the TiO_2 catalyst, and the UV lamp was located at the center of the reactor. A wire net at the inlet to the photoreactor created a laminar flow within the reactor. Irradiation was done with the following UV lamps: a BLB black light blue lamp with a maximum light intensity output at 365 nm (Sankyo Denki Co. Ltd., Japan), a GL4 germicidal lamp with a maximum at 254 nm (Sankyo Denki Co. Ltd., Japan), and a GLZ4 low-pressure mercury lamp with a maximum at 254 nm and a smaller (<5%) emission at 185 nm (Sankyo Denki Co. Ltd., Japan). The electric power consumption of all UV lamps was identical (4 W).

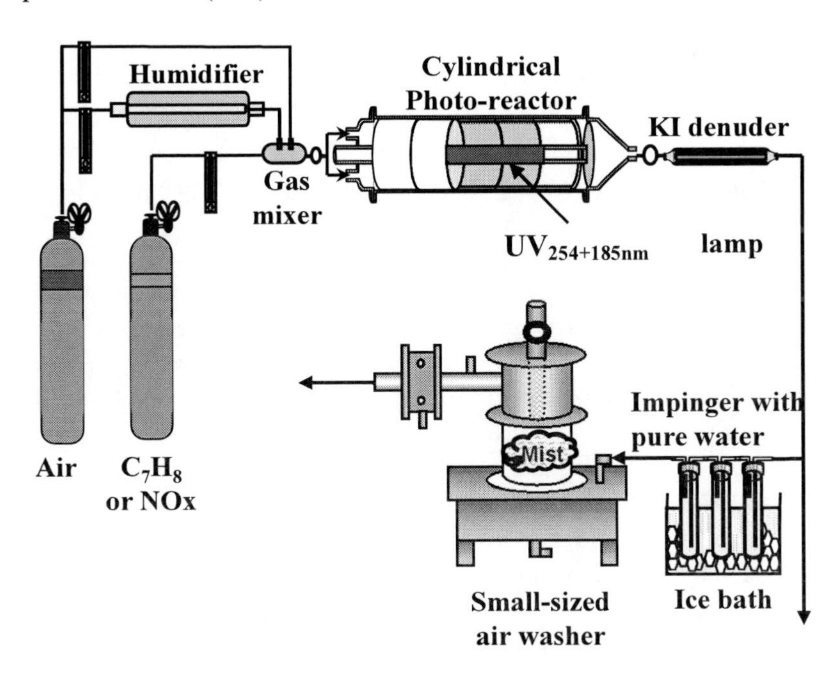

Figure 2. Schematic diagram of experimental apparatus.

2.1.2. Catalyst Preparation

The catalyst used were Degussa P-25 TiO_2 (75% anatase/ 25% rutile; Nippon Aerosil Co. Ltd.) with a BET surface area of 50 $m^2\ g^{-1}$. A sonicated mixture of 5 g of the TiO_2 catalyst in 300 ml of deionized water was dip-coated onto the innermost 3 cylinders of the photoreactor, and the cylinders were dried at 150 °C for 2 h in an oven. After the same procedure was repeated 4 more times, the cylinders were dried at 150 °C for 24 h. Consequently, approximately 40 mg of TiO_2 was coated on the inside of the cylinders (≈0.15 mg cm^{-2}). The ozone-decomposition catalyst (ODC; Nippon Syokubai Co. Ltd.) was a honeycomb-type catalyst consisting of a MnO_2 active element on a TiO_2/SiO_2 support. The reactor with ODC was located downstream the photoreactor. The used ODC volume was approximately 142 cm^3.

2.1.3. Experimental Procedure and Conditions

The experimental apparatus is illustrated in Figure 1(b). The desired toluene (the model VOC) concentration was adjusted by mixing toluene standard gas with clean air. Water vapor was obtained by passing dried air through a porous polytetrafluoroethylene tube containing deionized water at room temperature. The reaction temperature was maintained at room temperature (298 ± 1 K). Before the gas stream containing toluene and water vapor was introduced into the photoreactor, humidified air was allowed to flow through the illuminated photoreactor for several hours to pretreat the catalyst. The UV irradiation experiment was started after the inlet and outlet toluene concentrations were equal (1 h). All experiments were performed at least 3 times. The tested initial concentration of toluene was in the 0.6–10 ppmv range, and the relative humidity (RH) in most experiments was 40% (the RH was <1% in one set of experiments). Most experiments were performed at a relatively long residence time (33 s), to allow for a laminar flow and stable photodegradation.

2.1.4. Analytical Methods

The toluene concentration in the effluent gas was continuously monitored with a GC-FID gas chromatograph (GC 14B, Shimadzu Co. Ltd.) equipped with a gas autosampler. CO_2 and CO in the effluent gas were determined simultaneously with a GC-FID gas chromatograph (GC 15A, Shimadzu Co. Ltd.) equipped with a methane converter (MT-221, GL Science Co. Ltd.). DNPH-silica cartridges (Waters) were used for collection of aldehydes from effluent gas stream. The sample volume was 40 L at an air flow rate of 0.5–0.6

L min^{-1}. Aldehydes collected with the cartridge were extracted with acetonitrile and analyzed by a high performance liquid chromatograph (LC-9A, Shimadzu Co. Ltd.). In the $UV_{254+185nm}$ irradiation experiments, a KI (potassium iodide)-coated annular denuder was used to selectively remove the ozone, since effluent gas with a high level of ozone can seriously damage the GC column [26].

2.2. Application of an Air Washer with $UV_{254+185nm}/TiO_2$ Photocatalytic Reaction for VOC Degradation

2.2.1. Phtoreactor and UV Sources

Figure 2 shows a schematic diagram of the experimental apparatus. The cylindrical photoreactor ($V = 0.55$ L) has five inner cylinders.

In some of the reactions, a TiO_2 catalyst (Degussa P25) was used with UV irradiation to enhance the photodegradation. A sonicated mixture of 5 g of the TiO_2 catalyst in 300 mL of deionized water was dip-coated onto the innermost three cylinders of the photoreactor, and the cylinders were dried at 150 °C for 2 h in an oven. The dipping procedure was repeated four times, and then the cylinders were dried at 150 °C for 24 h. After the dipping procedure, the TiO_2 catalyst load on the insides of the cylinders was approximately 40 mg (*ca.* 0.15 mg cm^{-2}). The UV lamp was located at the center of the reactor. An ozone-producing low-pressure mercury lamp (GLZ4, Sankyo denki, Tokyo) with a maximum emission at 254 nm and a minor emission (*ca.* 5%) at 185 nm was used for irradiation. A laboratory-scale air washer with one water-spraying nozzle was used. Deionized water from the water tank (25 L) was sprayed (spray flow rate, 25 mL min^{-1}; spray pressure, 4 kg cm^{-1}) into the effluent gas stream emitted from the photoreactor. The mist was recirculated into the water tank.

2.2.2. Experimental Procedure and Conditions

The NOx or toluene gas was introduced into the photoreactor at a flow rate of 3 L min^{-1} and a relative humidity of ca. 40%. After the inlet and outlet concentrations reached equilibrium, the UV lamp was turned on. The removal ratios for the pollutants were calculated from the concentrations of the pollutants before and after $UV_{254+185nm}$ irradiation, and mass balance was established by quantifying the decomposition products. We measured the quantities of CO_2 and water-soluble organic intermediates (WSOI) obtained from toluene degradation and the quantities of NO_2 or HNO_3 obtained from

NOx degradation. The HNO_3 and WSOI selectivities were determined from the quantities collected in a series of three impingers containing pure water. The air-flow rate ranged from 0.8 to 1.0 L min^{-1} and the total sampled gas volume was 20 L for HNO_3 and 150 L for WSOI. The impingers were maintained at 5 °C to increase collection efficiency during sampling.

2.2.3. Analytical Methods

The toluene gas was monitored with a gas chromatograph (GC-14B, Shimadzu, Kyoto) equipped with a gas autosampler and a flame ionization detector. Carbon dioxide (CO_2) and carbon monoxide (CO) in the effluent gas were analyzed with a gas chromatograph (GC-15A, Shimadzu, Kyoto) equipped with a methane converter (MT-221, GL Science, Tokyo) and a flame ionization detector. A total organic compounds (TOC) analyzer (TOC-5000, Shimadzu, Kyoto) was used to quantify the total organic intermediates collected in the water. NO and NO_2 were separately monitored with a chemiluminescence NOx analyzer. The liquid samples obtained from the impingers, and the air washer were analyzed for HNO_2 and HNO_3 by means of an anion chromatograph (DX-100, Dionex, USA) equipped with an electroconductivity detector. The NOx analyzer could read HNO_3 as NO_2 because the analyzer contains a reduction catalyst capable of converting HNO_3 into NO. Thus, the NO_2 concentration obtained from the NOx analyzer was the sum of the NO_2 and HNO_3 concentrations. The NOx analyzer's conversion efficiency for HNO_3, obtained by means of a calibration test using standard gaseousHNO_3, was 85–90%. Ozone in the effluent gas was removed with a potassium iodide (KI)-coated annular denuder [26] because ozone could introduce experimental error as well as seriously damage the GC column.

2.3. Degradation of Organic Gases Using Ultrasonic Mist Generated from TiO_2 Suspension

2.3.1. Generation of Ultrasonic Mist Containing Photocatalyst Particles (UMP)

The experimental setup for degradation of organic gases with UMP is shown in Figure 3. The experimental reactor (*ca.* 2 L) consisting of Pyrex glass was equipped with an ultrasonic transducer (Honda Electronics, HM-303N, 2.4 MHz) as also shown in Figure 3. UMP containing titanium dioxide was generated inside the reactor when a TiO_2 suspension in Milli-Q water was irradiated with 2.4-MHz ultrasound. In general, photocatalytic reactions are based on the production of hydroxyl radical on the TiO_2 surface. A Degussa P-25 TiO_2 photocatalyst (Nippon Aerosil) was used in all experiments since P-25 TiO_2 particles can generate a sufficient amount of OH radicals, even in the liquid phase [27]. The crystal structure of the P-25 TiO_2 particles was approximately 80% anatase and 20% rutile, and the average particle diameter was about 30 nm. The surface area of TiO_2, as measured with a BET surface analyzer (Micromeritics, Flowsorb III-2305), was 50 m^2 g^{-1} [28,29]. The TiO_2 concentration was fixed at 1.0 g L^{-1}, which was the optimal condition found in our previous study [25]. The clean air that consists of pure N_2 and O_2 (4:1) was used as a carrier gas for all experiments. It barely contains other gas elements (H_2O < 10 ppm, CO_2 < 1 ppm) and particles. The generated UMP was cooled and trapped with a Dimroth condenser and returned into the reactor. Three types of low-pressure mercury UV lamps with identical electronic power consumption (4 W) and differing wavelengths were used as light sources in this study: a BLB black light blue lamp with a maximum light intensity output at 365 nm (Sankyo Denki, FL4BLB), a short wavelength UV germicidal lamp with a maximum at 254 nm (Sankyo Denki, GL4), and ozone lamp with a maximum at 254 nm and a smaller (3%) emission at 185 nm (Sankyo Denki, GL4ZH). The detailed emission spectra of the lamps are shown elsewhere [30]. Under UV irradiation, the relative humidity was measured using a hygro-thermometer (Hanna, HI-8564) at the reactor exit for all conditions since the relative humidity affected the stability of the mist. The observed relative humidity was over 95%, which was outside of the measurement range for all conditions, indicating that the mist in the reactor could be maintained in a stable condition.

2.3.2. Degradation of Organic Gases

Toluene, *p*-xylene, styrene (highly hydrophobic), formaldehyde and acetaldehyde (highly hydrophilic) were used as experimental gases due to their differing properties. Toluene, *p*-xylene, and styrene were introduced into the reactor at 1.0 L min^{-1} after dilution of the standard gas. Formaldehyde and acetaldehyde gases were generated from paraformaldehyde or pure acetaldehyde by a diffusion method in a thermostatic bath (60 °C) and an ice bath with an impinger (2 °C) respectively, and introduced into the reactor at 1.0 L min^{-1}. The concentration of the introduced pollutant gas was 0.6 ppm for the removal of toluene, while 4.0 ppm of all pollutant gases including toluene was introduced into the reactor to accurately estimate the mineralization of the various gas species. After the introduction of the experimental gases, UMP was generated under UV lamp irradiation and the temperature inside the reactor was maintained at 25 ± 5 °C by cooling with a water bath. In this reactor, the UMP did not stay on the surface of the UV lamp for a long time because the surface of the UV lamp was smooth and the relative humidity was so high (>95%). Therefore, the decrease in reaction ratio due to shading light was not confirmed in the experiment.

2.3.3. Analytical Methods

The VOC and aldehyde gas concentrations were determined with a gas chromatograph equipped with flame ionization detection (GC-FID; Shimadzu, GC-7A) and a high-performance liquid chromatograph equipped with UV detection (HPLC-UV; Shimadzu, LC-9A), respectively. Prior to HPLC-UV analysis, formaldehyde and acetaldehyde were derivatized with 2,4-dinitrophenylhydrazine. The mineralized decomposition products (CO and CO_2) were measured by GC-FID with a methanizer (GL Sciences, MT-221). To detect intermediates formed by photocatalytic decomposition of toluene in the UMP, the UMP was collected at the outlet of the reactor by bubbling through a Milli-Q water-filled impinger, and total carbon in the collected UMP was measured with a TOC analyzer (Shimadzu, TOC-V). Ozone was formed in the experiment using $UV_{254+185nm}$ irradiation. Since a high concentration of O_3 (ca 52.0 ppm) was observed at the reactor exit under $UV_{254+185nm}$ irradiation, O_3 was diluted with air and measured with a UV-absorption ozone analyzer (Shimadzu, UVAD-1000). In the degradation experiments for organic gases, O_3 was removed by a potassium iodide denuder prior to injection of the VOC gases into the GC-FID because O_3 causes serious damage to GC columns. Furthermore, the droplet size distribution of UMP and the fine particles composed of decomposition intermediates under $UV_{254+185nm}$ irradiation were measured at the reactor outlet with a scanning mobility particle sizer (SMPS; TSI, Model-3934) for fine droplets and an optical particle counter (OPC, TSI, Model-8240) for coarse droplets, respectively. An electric furnace was connected to the particle counters (SMPS and OPC), and set at 200 °C in order to measure the diameter of the TiO_2 particles in the UMP. The removal efficiency was obtained based on upstream and downstream concentration, and the mineralization ratio was calculated using the following equation:

$$\text{Mineralization ratio (\%)} = \frac{C_{CO_2,t} + C_{CO,t}}{N_C(C_0 - C_t)} \times 100 \quad (1)$$

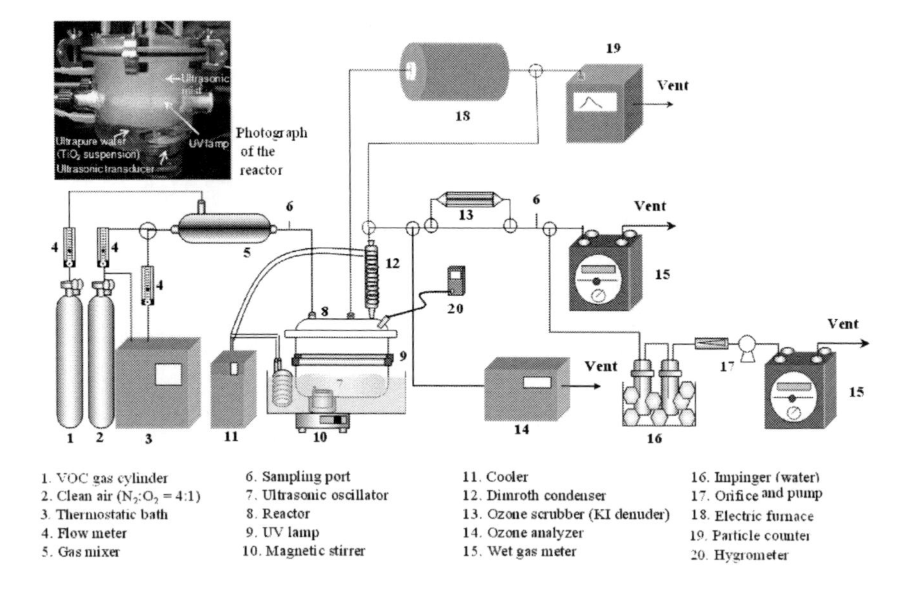

Figure 3. Schematic diagram of experimental setup for degradation of VOC gases by UMP.

Here, C_0 is the upstream initial concentration, C_t is the downstream concentration after time t, $C_{CO2,t}$ and $C_{CO,t}$ are the CO_2 and CO concentrations at time t, and N_C is the carbon number of the supplied organic gas.

3. Results and Discussions

3.1. Application of $UV_{254+185nm}$ Irradiation to the VOC Degradation using TiO_2 Photocatalyst

3.1.1. Photochemical and Photocatalytic Degradation of Gaseous Toluene by TiO_2 Catalyst with $UV_{254+185nm}$ Irradiation

To characterize the photochemical and photocatalytic reaction, we examined photodegradation of gaseous toluene in the presence or absence of TiO_2 under $UV_{254+185nm}$ irradiation, and the experiments were carried out in two reaction media, such as nitrogen stream (no photooxidation of toluene) and air stream (photooxidation of toluene). Approximately 10% conversion was obtained without TiO_2 in the dry nitrogen stream, indicating the photolysis of toluene upon irradiation with $UV_{254+185\ nm}$ irradiation (see Figure 4).

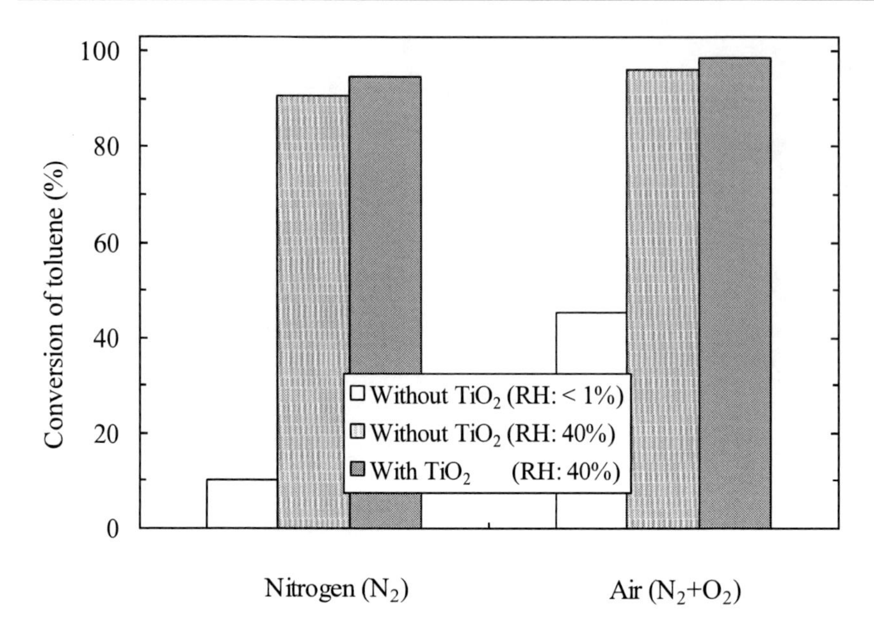

Figure 4. Effect of reaction medium on the conversion of toluene with $UV_{254+185nm}$ irradiation in the presence or absence of TiO_2 catalyst. $[C_7H_8]_0$: 2 ppmv, residence time: 33 s, relative humidity: ≈ <1%, 40%.

Since nitrogen does not absorb light above 125 nm, and no oxidative reaction occurs under that condition, the degradation of organic compounds in a dry nitrogen stream should be due to photolysis only [20). When water vapor]was added to the dry nitrogen stream, the conversion increased dramatically from 10% to 91%, even without TiO_2 catalyst; with TiO_2 catalyst, the conversion was further enhanced. Since water molecules exhibit a continuous UV adsorption spectrum between 175 and 190 nm, a plenty of OH radicals can be produced in the gas phase (Eq. (2)) [19]. Thus, we attribute the dramatic increase of conversion in the humid nitrogen stream to abundant OH radicals formed in the gas phase. Reaction in a nitrogen stream, however, is not recommended, because complete oxidation to CO_2 is not achieved in the absence of oxygen.

$$H_2O + h\nu \rightarrow H + OH^{\cdot} \quad (2)$$

Meanwhile, a conversion of approximately 46% was obtained in the dry air stream without TiO_2 catalyst, and toluene was also drastically decomposed in the humid air stream, even without TiO_2 catalyst. A nearly complete conversion of toluene was achieved in the presence of TiO_2 catalyst. With the dried air stream, reactive species of oxygen such as O_3, $O(^1D)$ and $O(^3P)$ may be formed in the reactor upon $UV_{254+185nm}$ irradiation (Eqs. (3)–(5)); conversely, formation of OH radicals is limited. Thus, it is reasonable to assume that the reactive species of oxygen also had an important effect on the decomposition of toluene, since a conversion of 46% was obtained with the dried air stream.

$$O_2 + h\nu\ (<243\ \text{nm}) \rightarrow O(^1D) + O(^3P) \quad (3)$$

$$O(^1D) + M \rightarrow O(^3P) + M\ (M = O_2 \text{ or } N_2) \quad (4)$$

$$O(^3P) + O_2 + M \rightarrow O_3 + M \quad (5)$$

$$O_3 + h\nu\ (<310\ \text{nm}) \rightarrow O(^1D) + O_2 \quad (6)$$

$$O(^1D) + H_2O \rightarrow 2{\cdot}OH \quad (7)$$

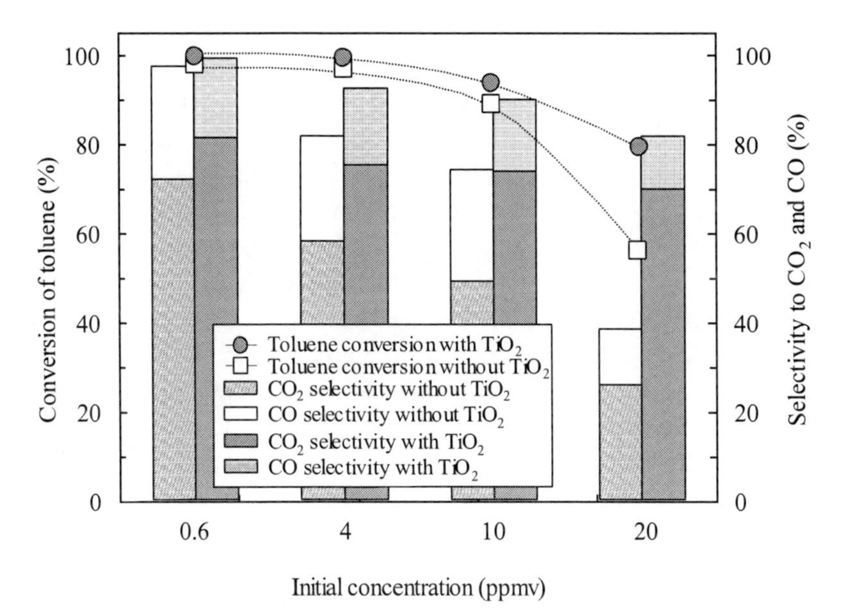

Figure 5. Effect of initial concentration on the conversion and mineralization of toluene with $UV_{254+185nm}$ irradiation in the presence or absence of TiO_2 catalyst. Residence time: 33 s, relative humidity: ≈ 40%.

Table 1. Conversion of toluene and selectivity to decomposition products with irradiation of three UV sources in the presence of TiO_2 catalyst

UV wavelength with TiO_2	Conversion (%)	Selectivity		
		CO_2	CO	HCHO
UV_{365nm}/TiO_2	83	62.1	12.4	0.57
UV_{254nm}/TiO_2	84	73.4	14.6	0.97
$UV_{254+185nm}/TiO_2$	99	80.7	18.6	ND[a]

$[C_7H_8]_0$: 0.6 ppmv, residence time: 33 s, relative humidity: ≈ 40%. [a]ND: not detected.

In addition, since adding water vapor in an air stream can also yield OH radicals to react with electronically excited reactive species of oxygen atoms ($O(^1D)$) produced with the photolysis of ozone (Eqs. (6) and (7)) [9,19], abundant reactive species including OH radicals will exist in the reactor. Effective decomposition of toluene without TiO_2 catalyst in humid air stream was due to photochemical degradation by abundant reactive species in gas phase. On the other hand, the use of TiO_2 catalyst under $UV_{254+185nm}$ irradiation could increase the conversion of toluene regardless of the reaction medium. When the TiO_2 catalyst absorbs photons with energy exceeding the band gap energy, photocatalytic oxidation initiated from electron–hole pairs on the surface of the TiO_2 catalyst has been well characterized [14]. Therefore, we attributed the increased conversion by the use of TiO_2 under $UV_{254+185nm}$ irradiation to the combination effect of photochemical oxidation in the gas phase and photocatalytic oxidation over the TiO_2 catalyst.

In order to corroborate the effect of combined photochemical and photocatalytic oxidation, we investigated the conversion and mineralization of toluene with variation of the initial concentration in the presence and absence of TiO_2 under $UV_{254+185nm}$ irradiation. The results showed that high conversion was observed in the initial concentrations range from 0.6 to 10 ppmv, regardless the presence of TiO_2 (See Figure 5).

However, in that concentration range, the mineralization behavior observed differed somewhat from the conversion behavior. The mineralization decreased rapidly with the rise of initial concentration in the absence of TiO_2. In presence of the TiO_2, however, a high degree of mineralization, ≈ 90%, was obtained even at 10 ppmv, which confirms the photocatalytic role of TiO_2 catalyst. Furthermore, the notable effect of TiO_2 was clearer at 20 ppmv. These observations also suggest that $UV_{254+185nm}$ irradiation can efficiently excite the TiO_2. We suppose that the gaseous toluene is photochemically degraded by abundant reactive species including hydroxyl radicals in the gas phase, the residual toluene and its intermediates are then moved onto the photoirradiated TiO_2 surface, and afterwards photocatalytically mineralized.

3.1.2. Photodegradation of Gaseous Toluene by Irradiation of Different UV Sources in Presence of TiO_2 Catalyst

1) Comparison of Conversion, Mineralization and Generation of Harmful Intermediates

Gaseous toluene was subjected to UV_{365nm}, UV_{254nm}, and $UV_{254+185nm}$ irradiation in the presence of TiO_2 catalyst. The conversion of toluene and selectivity to decomposition products under the three UV irradiation conditions is shown in Table 1. The result shows that the amount of toluene converted under $UV_{254+185nm}$ irradiation was much higher than that converted under UV_{365nm} or UV_{254nm} irradiation. For $UV_{254+185nm}$ irradiation, conversions of 99% were obtained with TiO_2 catalyst, whereas conversions of no more than 83% and 84% were obtained for UV_{254nm} and UV_{365nm} irradiation, respectively.

In addition, the selectivity to CO_2 and CO, which is expressed in terms of mineralization, was enhanced, and emission of formaldehyde, which is known as a noxious chemical, was inhibited significantly under $UV_{254+185nm}$ irradiation. Formation of formaldehyde from photocatalytic degradegradation of toluene with UV_{254nm} and UV_{365nm} irradiation has been reported elsewhere [31,32]. We attribute the remarkable improvement of photodegradation with $UV_{254+185nm}$ irradiation to a combination effect of photochemical degradation in the gas phase and photocatalytic degradation over the TiO_2 catalyst as mentioned earlier.

2) Comparison of the Effect of Initial Concentration

The effects of initial concentration on the photodegradation of toluene with the three UV sources are illustrated in Figure 6. With an initial concentration of 0.6 or 2 ppmv and irradiation at UV_{365nm} or UV_{254nm}, conversions of approximately 83–84% were obtained, except in one case (2 ppmv initial concentration, UV_{365nm} irradiation), in which a slightly lower conversion was obtained. With an initial concentration of 10 ppmv and irradiation at UV_{365nm} or UV_{254nm}, conversion dropped rapidly with irradiation time, and the color of the catalyst changed from white to yellow. With $UV_{254+185nm}$ irradiation, however, a conversion above 95% was obtained for all tested concentrations, with no drop in conversion efficiency over time.

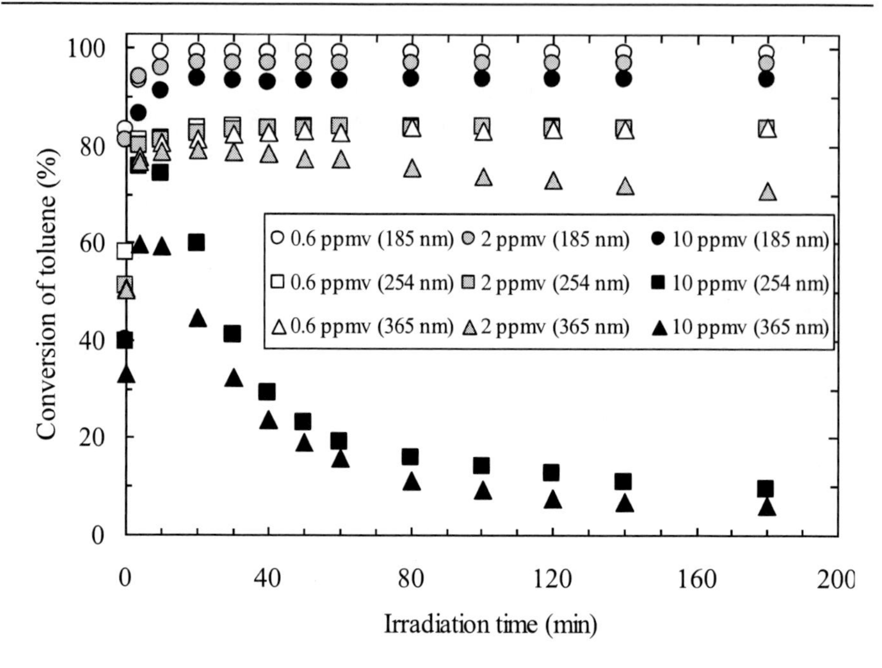

Figure 6. Effect of initial concentration on the conversion of toluene with the three UV light sources irradiation in the presence of TiO_2 catalyst. $[C_7H_8]0$: 0.6, 2, 10 ppmv, residence time: 33 s, relative humidity: ≈ 40%.

The reaction rate over a TiO_2 catalyst at a low flow rate depends on only the bulk fluid-to-catalyst diffusion mass transport, because intraparticle mass transport is negligible, owed to the nonporous nature of the thin-film catalyst [1,31-33]. Of course, adsorption of the reactants onto the catalyst surface is also important, but such adsorption is probably not rate limiting if sufficient active sites are available. These facts may explain why conversion was independent of concentration in the 0.6–2 ppmv range throughout the UV_{254nm} irradiation experiments and in the initial stage of the UV_{365nm} irradiation experiment. In addition, these experiments were carried out in a laminar flow at a low flow rate, and sufficient active catalytic sites were available for reaction with toluene.

With $UV_{254+185nm}$ irradiation, conversions above 93% regardless of the initial concentration indicate that abundant reactive species were formed in the gas phase including the surface of the TiO_2 catalyst for reaction with toluene. It is important to note that serious catalyst deactivation occurred with the UV_{365nm} and UV_{254nm} light sources when the initial concentration was 10 ppmv. Catalyst deactivation is a substantial obstacle to the industrial application of photocatalytic degradation. However, the use of $UV_{254+185nm}$

irradiation could significantly avoid catalyst deactivation at high concentration.

3) Comparison of Catalyst Deactivation and Regeneration

The deactivation and regeneration characteristics of the TiO2 catalyst were investigated for each of the UV sources. Figure 7 shows the variation for the conversion of toluene as the initial concentration at the inlet was varied in the order 0.6 ppmv → 4 ppmv → 0.6 ppmv.

Initially, with UV_{365nm} and UV_{254nm} lights and an inlet toluene concentration of 0.6 ppmv, no catalyst deactivation was observed for 180 minutes. When 4 ppmv of toluene was introduced, however, the conversion ratio gradually decreased with increasing irradiation time, and the color of the TiO_2 catalyst changed from white to yellow, indicating that intermediates, such as benzaldehyde, benzyl alcohol, and benzoic acid, were deposited on the TiO_2 catalyst [6]. When 0.6 ppmv of toluene was introduced again, the conversion ratio began to steadily increase, and after photoirradiation for approximately 2–3 h, the original conversion ratio was obtained. In contrast, with $UV_{254+185nm}$ irradiation, a high conversion ratio was maintained throughout the tested concentration range, without any catalyst deactivation, although there was a slight drop in the conversion ratio at 4 ppmv. The TiO_2 catalyst did not change color during irradiation.

Serious deactivation of the TiO_2 catalyst during PCO of aromatic compounds at high concentration has been commonly reported [4,5]. But, Ao et al. [3] and Jo et al. [34] reported that deactivation is unlikely with trace levels of pollutants.

At 0.6 ppmv with UV_{365nm} or UV_{254nm} irradiation, the less reactive intermediates are probably oxidized faster than they are deposited on the TiO_2 catalyst surface, which is likely why the catalyst is not deactivated under these conditions. While, the serious deactivation at 4 ppmv indicates that a temporary influx of a high concentration of organic compounds into the photoreactor may cause serious problems in case of some manufacturing processes. The deactivated catalyst would eventually be regenerated by a self-cleaning effect if the concentration returned to a low level (Figure 7(c)). However, complete restoration of the catalyst would take a long time, and it might be necessary to prepare the preparatory purification facilities. In this regard, the use of $UV_{254+185nm}$ light is a very effective means for avoiding deactivation of the TiO_2 catalyst.

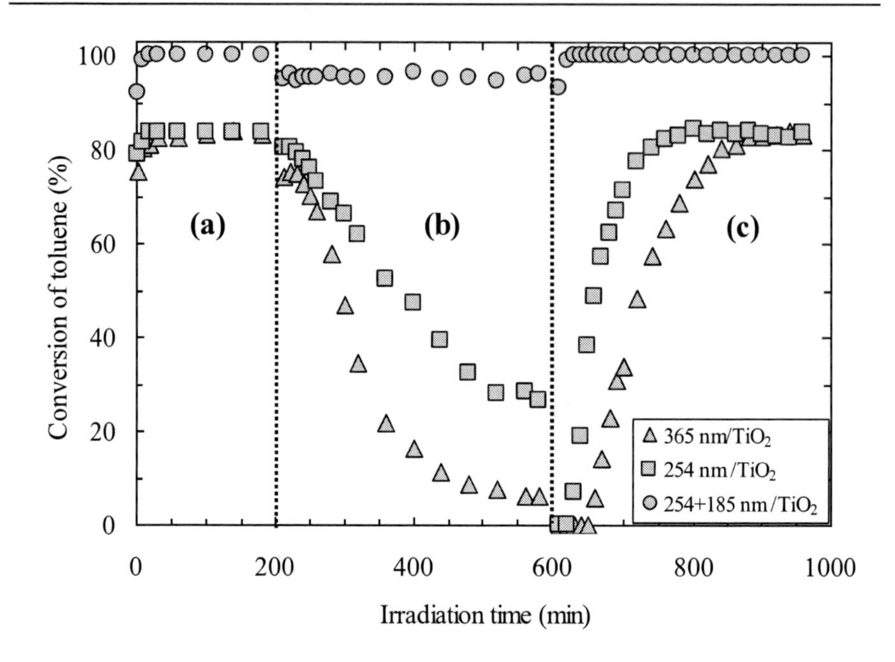

Figure 7. Time course for conversion of toluene with continuous variation of the concentration at the inlet for each UV source. (a) $[C_7H_8]_0$: 0.6 ppmv, (b) $[C_7H_8]_0$: 4 ppmv, (c) $[C_7H_8]_0$: 0.6 ppmv, residence time: 33 s, relative humidity: ≈40%.

The deactivation–regeneration process was investigated in detail. First, the TiO_2 catalyst was exposed to 10 ppmv of toluene for 2 h under UV_{365nm} irradiation; this treatment deactivated the catalyst (Figure 8(a)). The 365 nm UV lamp was then replaced with the $UV_{254+185nm}$ lamp, and the catalyst was irradiated for 30 min in the absence of toluene to regenerate the catalyst (Figure 8(b)). The lamps were then interchanged, and 10 ppmv of toluene was irradiated at UV_{365nm} as before to determine the degree of regeneration of the catalyst (Figure 8(c)). The experiments were carried out for several regeneration conditions.

The degree of regeneration in humid nitrogen stream with $UV_{254+185nm}$ irradiation was low and similar to the degree of regeneration in a humid air stream without UV254+185nm irradiation. A dried air stream with UV254+185nm irradiation led to a high degree of regeneration. In addition, in the absence of UV irradiation, externally introduced ozone effectively decomposed the intermediates over the TiO_2 catalyst. With humid air and $UV_{254+185nm}$ irradiation, the conversion ratio returned to its original level, and the TiO_2 surface restored its original color.

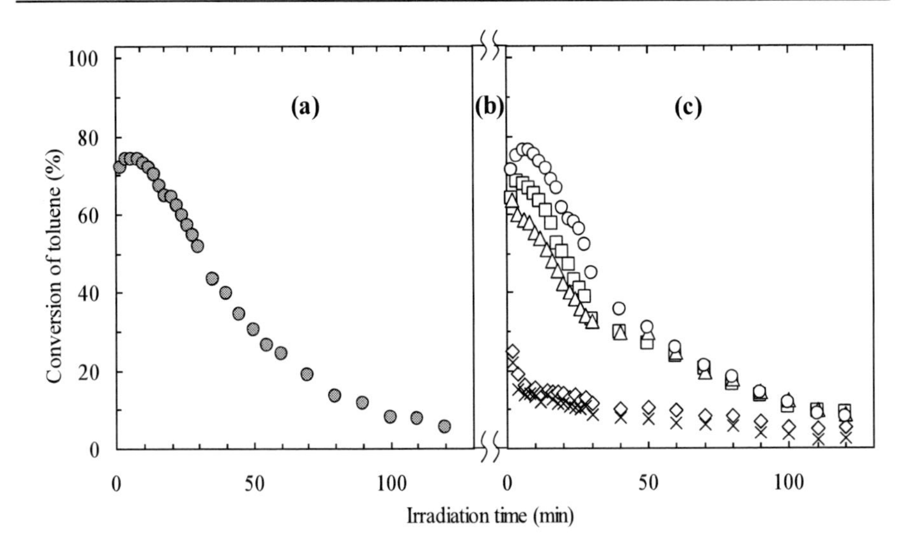

Figure 8. Deactivation and regeneration characteristics of the TiO_2 catalyst. (a) Photodegradation of toluene under UV_{365nm} irradiation with TiO_2 catalyst ($[C_7H_8]_0$: 10 ppmv, residence time: 47 s, relative humidity: ≈ 40%). (b) Regeneration treatment with $UV_{254+185nm}$ irradiation at several regeneration conditions without toluene feeding (regeneration time: 30 min, residence time: 33 s). (c) Photodegradation of toluene under UV_{365nm} irradiation with TiO_2 catalyst after each regeneration treatment (UV turn on + humidified air feeding (○), UV turn on + dried air feeding (□), UV turn off + dried air containing O_3 (≈ 27 ppm) feeding (△), UV turn on + humidified N_2 gas feeding (◇), UV turn off + humidified air feeding (×)) ($[C_7H_8]_0$: 10 ppmv, residence time: 47 s, relative humidity: ≈ 40%).

Ameen and Raupp [4] and Einaga et al. [8] suggested that OH radicals over the regenerated TiO_2 catalyst were the key factor in decomposing the carbon deposits on the catalyst. However, it seems that OH radicals alone cannot lead to complete decomposition of the carbon deposits on the TiO_2 catalyst, since little regeneration was observed in the humid nitrogen stream, where the existence of the numerous OH radicals is expected (Eq. (2)). In contrast, reactive species of oxygen, such as O^-_2, $O(^1D)$, and O_3, were indispensable for the complete oxidation of the carbon deposits to CO_2, since remarkable regeneration was observed when these reactive species were present. Therefore, we propose that complete regeneration in the humid air stream with $UV_{254+185nm}$ irradiation was due to effective decomposition of carbon deposits on the catalyst by reactive species of oxygen, OH radicals, and direct photolysis.

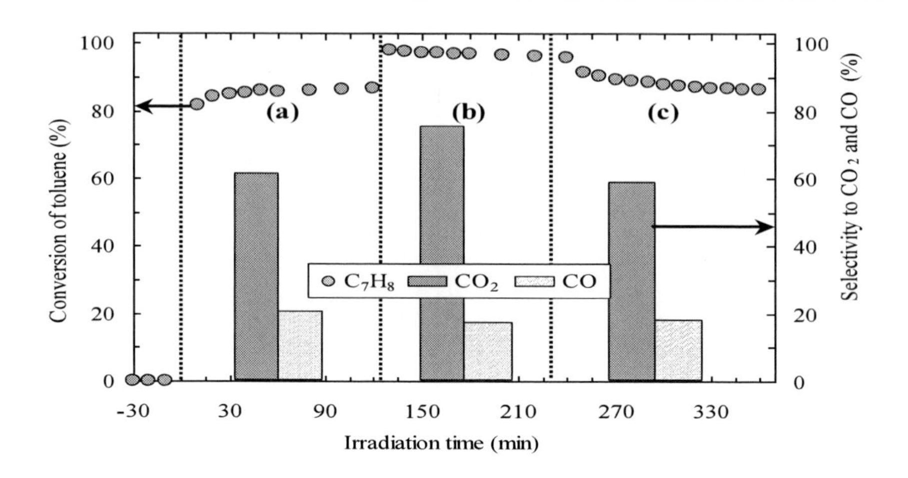

Figure 9. Variation of the toluene conversion and selectivity to CO_2 and CO by coupling the photoreactor and the ODC reactor. (a) Photoreactor alone, (b) photoreactor + ODC reactor (in presence of O_3), (c) photoreactor + ODC reactor (in absence of O_3). $[C_7H_8]_0$: 2 ppmv, residence time: 16.5 s, relative humidity: ≈ 40%, space velocity in the ODC layer: 875 h^{-1}.

These results strongly suggest that $UV_{254+185nm}$ irradiation can be successfully employed in the TiO_2-catalyzed treatment of aromatic VOCs, overcoming the problem of catalyst deactivation in classical photocatalytic degradation.

3.1.3. Formation and Treatment of By-products

Although photogenerated ozone enhanced the evolution of intermediates and prevented deactivation of the catalyst during the $UV_{254+185nm}$ irradiation, excess ozone in the effluent gas stream should be reduced to a safe level since it is a health hazard. Therefore, a reactor containing an ODC was set up behind the photoreactor to treat excess ozone in the effluent gas stream. The simultaneous elimination of some VOCs and ozone with this ODC has been described [21,35,36]. Thus, the conversion of both residual toluene and O_3 in the effluent gas by the ODC was investigated.

Figure 9 shows the variation for the conversion of toluene and selectivity to CO_2 and CO when a toluene gas stream was passed continuously from the photoreactor to the ODC reactor. The experiment was carried out in both the absence and presence of excess ozone in the effluent gas stream, to determine the role of ozone in the ODC layer. In a preliminary experiment, the conversion of toluene dropped quickly above a RH of approximately 40%, indicating that water molecules compete with organic molecules for the active sites of the ODC layer [21]. Therefore, a RH of approximately 40% was used in this study. The conversion of toluene was significantly enhanced when the effluent gas containing O_3 was passed to ODC reactor (Figure 9(b)). However, when the ozone was selectively removed from the effluent gas stream with a KI coated denuder, the conversion was diminished (Figure 9(c)), indicating that the excess ozone had a key role in the removal of toluene over the ODC layer. Moreover, the CO_2 yield was increased when the effluent gas containing O_3 was passed. This result demonstrates that intermediates were also effectively evolved. To explain these observations, we propose that triplet oxygen atoms ($O(^3P)$), formed during ozone-decomposition over the ODC layer, act as a mono-oxygenating agent toward toluene and its intermediates, although we have not yet established a detailed mechanism. Futamura et al. [37] experimentally demonstrated the formation of $O(^3P)$ from O_3 on the surface of MnO_2 in a silent discharge plasma reactor. Since the active component of the ODC is MnO_2, the same mechanism could occur here. This suggestion is supported by our previous report [21, 36].

Meanwhile, the excess ozone in the effluent gas was decomposed completely at ozone concentrations ranging from approximately 27 ppmv to below 0.05 ppmv, the limit recommended by WHO for ozone in indoor air. Accordingly, it is clear that the ODC is capable of simultaneously eliminating some organic compounds and excess ozone from an effluent gas stream. Although many ODCs, including MnO2, are affected by severe conditions such as a large space velocity and a high moisture level [38], using a mixed ODC/TiO_2 catalyst would minimize the effect of moisture, and employing a catalyst with a honeycomb structure would enhance treatment efficiency at higher space velocities [36].

It has been reported that secondary organic aerosols (SOA) are formed by gas-to-particle conversion of organic compounds in air upon irradiation at $UV_{254 + 185nm}$ [17]. Generally, the SOA are formed only from the oxidation of organic molecules sufficiently large to lead to products that have vapor pressures low enough to enable them to condense into the aerosol phase [39]. No SOA were observed in photocatalysis using UV_{365nm} or UV_{254nm} irradiation. Whereas, it was found that using $UV_{254+185nm}$ irradiation formed trace SOA, depending on the variation of the flow rate, RH, and initial concentration. The total volume concentration of SOA, formed from degradation of toluene in the presence and absence of TiO_2 under $UV_{254+185nm}$ irradiation, ranged from 1.53×10^{11} to 6.66×10^{10} $nm^3\ cm^{-3}$, respectively, at initial concentration of 4.0 ppmv, RH of $\approx$ 40% and RT of 33 s. The medium diameter was around 100 nm. It is important to note that the generated SOA concentration decreased in presence of TiO_2 catalyst, which indicates that an efficient application of the TiO_2 catalyst may adequately control the generation of them. In addition, the SOA can easily be removed by a classical purification method, such as the photoelectron method [40], an air filter, or a wet scrubber [41] compared to gaseous intermediates.

3.2. Application of an Air Washer with $UV_{254+185nm}/TiO_2$ Photocatalytic Reaction for VOC Degradation

3.2.1. Trapping Efficiency of the Air Washer

In a preliminary test, the trapping efficiencies of the air washer (AW) for water-insoluble NO, NO_2, toluene, and water-soluble HNO_3 were examined. The water-soluble HNO_3 was completely trapped (>99%) at the concentration range and flow rates tested, owing to its high water solubility. In contrast, less than 5% of the water-insoluble NO, NO_2, and toluene were trapped. Therefore, the water-insoluble pollutants had to be converted into water soluble materials by means of a pre-treatment procedure in order to be removed by the AW.

3.2.2. Photodegradation of NOx and Toluene with $UV_{254+185nm}$ Irradiation and the Water Solubility of Decomposition Products

$UV_{254+185\ nm}$ completely oxidized NO gas to HNO_3 (>*ca.* 90%) and NO_2 (<*ca.* 10%) (Table 1). NO_2 gas was almost completely oxidized (*ca.* 90%) by irradiation and was converted almost completely to HNO_3. $UV_{254+185nm}$ irradiation of humid air simultaneously produces abundant O_3 and OH radicals via photochemical reaction. Upon entering the UV-irradiated air, NO reacts rapidly with O_3 to formNO$_2$ (Eq. (8)), and thenNO$_2$ is readily oxidized to HNO_3 in the presence of OH radicals (Eq. (9)). In addition, NO_2 gas also reacts directly with OH radicals.

$$NO + O_3 \rightarrow NO_2 + O_2 \tag{8}$$

$$NO_2 + OH + M \rightarrow HNO_3 + M \ (M = O_2 \text{ or } N_2) \tag{9}$$

Therefore, it is reasonable to expect the combination of ozone and OH radicals to be very useful for converting NOx into HNO_3.

Table 2. Removal ratios of NOx and toluene and product selectivity after UV irradiation

	Removal ratio (%)	Product selectivity (%)
NO	100	HNO_3 (91), NO_2 (9)
NO_2	92	HNO_3 (100)
C_7H_8	60	CO_2 (62), CO (8), WSOI[b] (25), Unknown (5)

[a] $[NOx]_0$, $[C_7H_8]_0$ = *ca.* 0.6 ppm; RT, 11 s (3 L min^{-1}); RH, *ca.* 40%.
[b] WSOI, water-soluble organic intermediates.

Methods that can simultaneously produce large quantities of O_3 and OH radicals are rare. A corona discharge technique was recently used to effectively produce ozone and OH radicals, but the energy efficiency of the technique is low, and undesirable by-products are produced [42].

The removal ratio for the photodegradation of toluene was moderate compared to that for NOx, owing to toluene's lower reactivity with OH radicals. Nevertheless, the organic intermediates that formed were water-soluble compounds that could be collected efficiently with water, which implies that the intermediates can be washed out by a gas washing technique, such as an air washer.

3.2.3. Treatment of NOx and Toluene with the $UV_{254+185nm}$ Irradiation System and the Air Washer

When NO and NO_2 were subjected to $UV_{254+185nm}$ irradiation, the NO concentration dropped rapidly to zero and the NO_2 concentration increased (Figure 10). As the reaction proceeded, the NO_2 concentration slowly increased, reaching a steady state after 400 min (7 h). The time course of the NO_2 concentration is thought to arise from the fact that the NOx analyzer counts HNO_3 as NO_2. In this initial stage, HNO_3 is thought to be strongly adsorbed on the wall of the photoreactor and venting line, so the NO_2 value was low. As the adsorption of HNO_3 decreased, the NO_2 concentration increased, and when the adsorption–desorption of HNO_3 reached equilibrium, the NO_2 value stabilized at its highest value. Note, however, that in this equilibrium state, the NO_2 value did not reach the initial concentration of NO. This result can be explained by the reduction ratio of HNO_3 in the NOx analyzer.

When the effluent gas containing HNO_3 was passed through the AW, the NO_2 concentration rapidly dropped and reached a new, lower equilibrium value, since HNO_3 was trapped in the AW.

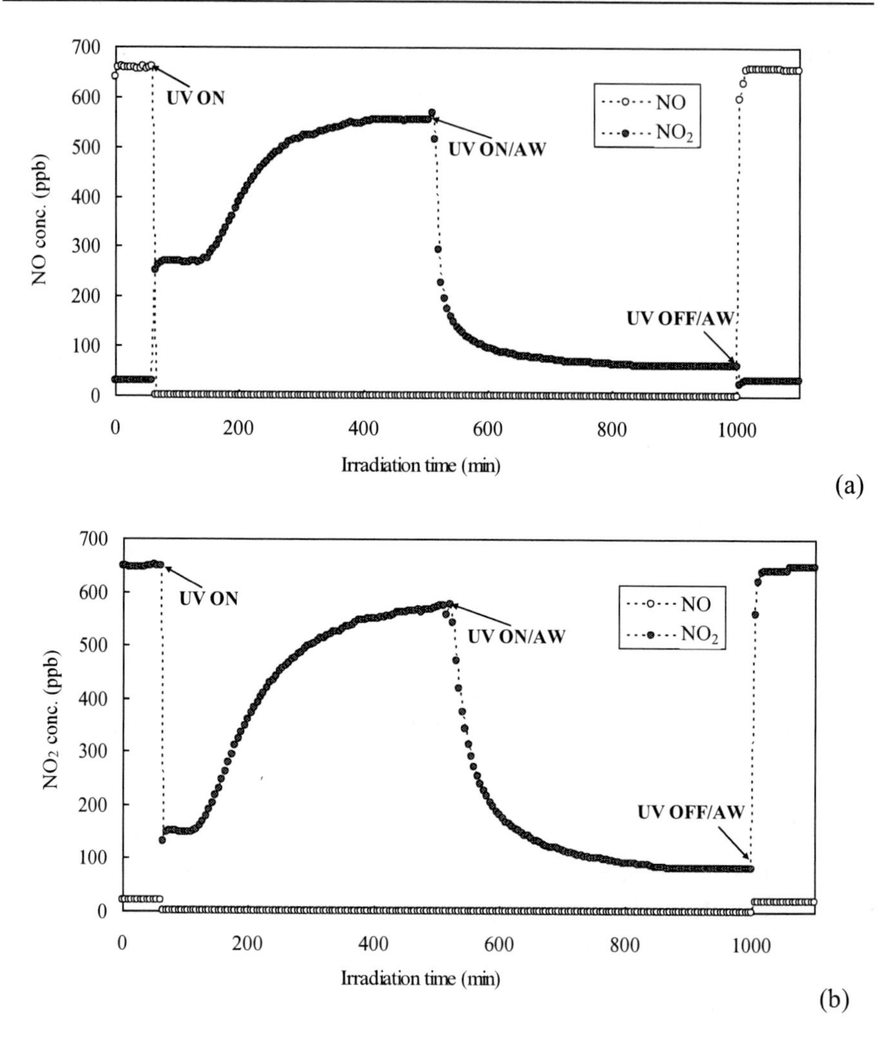

Figure 10. (a) Time course of NO concentration upon treatment with UV irradiation and the AW. $[NO]_0$, *ca*. 0.65 ppm; RT, 11 s (3 L min^{-1}); RH, *ca*. 40%. (b) Time course of NO_2 concentration upon treatment with UV irradiation and the AW. $[NO_2]_0$, ca. 0.65 ppm; RT, 11 s (3 L min^{-1}); RH, *ca*. 40%.

The time course of the photodegradation of NO_2 was similar to that of NO. To confirm the trapping ability of the AW, we continuously analyzed the HNO_3 concentration in the trap water during the operation (Figure 11) and found that the HNO_3 concentration in the water increased with time, indicating that HNO_3 was actually trapped by the mist.

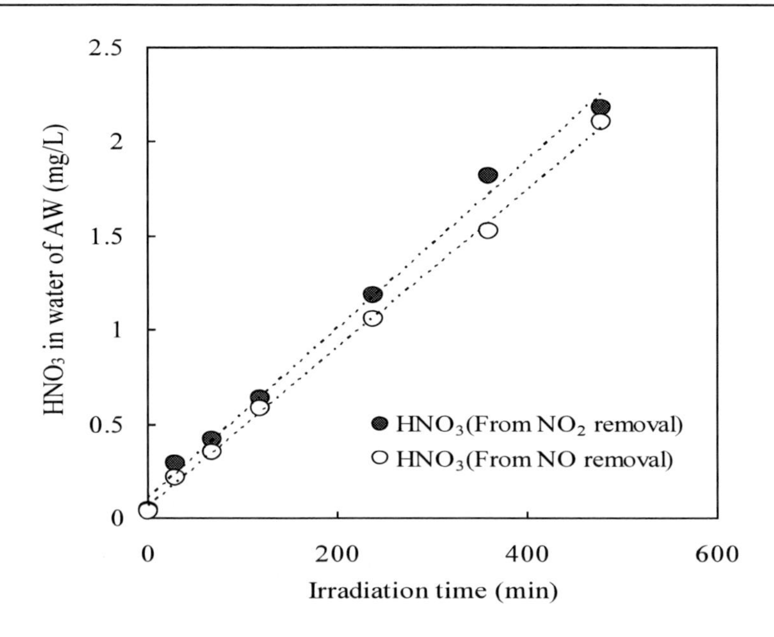

Figure 11. Time dependence of the HNO_3 concentration in the trap water of the AW.

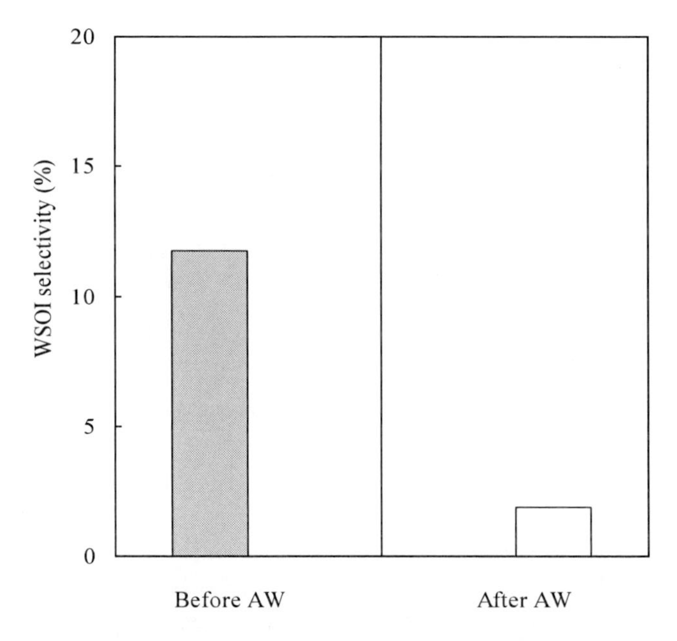

Figure 12. WSOI selectivity of toluene decomposition, before and after passage through the AW. $[C_7H_8]_0$, 0.6 ppm; RH, *ca*. 40%; RT, 33 s (3 L min^{-1}).

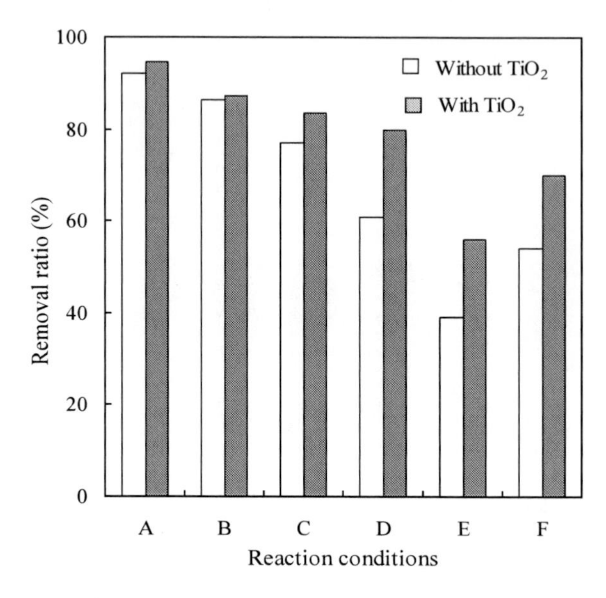

Figure 13. Effect of TiO_2 on the $UV_{254+185\ nm}$ irradiation of NO_2 and toluene. A: $[NO_2]0$, 0.6 ppm; RT, 11 s (3 L min^{-1}). B: $[NO_2]0$, 10 ppm; RT, 11 s (3 L min^{-1}). C: $[NO_2]0$, 0.6 ppm; RT, 6.6 s (5 L min^{-1}). D: $[C_7H_8]_0$, 0.6 ppm; RT, 11 s (3 L min^{-1}). E: $[C_7H_8]_0$, 10 ppm; RT, 11 s (3 L min^{-1}). F: $[C_7H_8]_0$, 0.6 ppm;RT, 6.6 s (5 L min^{-1}). RH, *ca.* 40%.

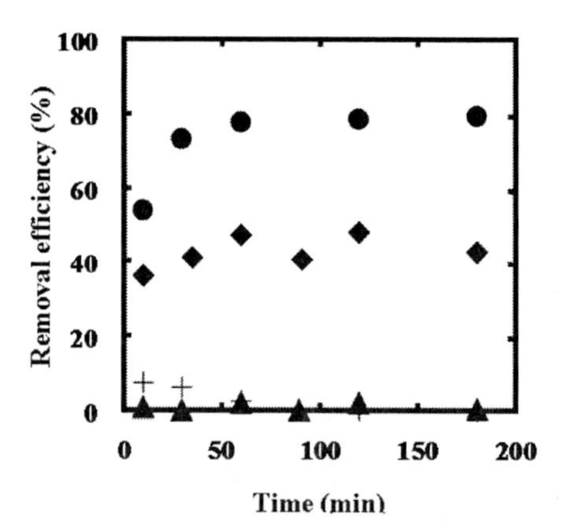

Figure 14. Comparison of toluene removal efficiencies for the following reaction conditions: (●) US/UV_{365nm}/TiO_2, (△) US, (◆) UV_{365nm}/TiO_2, and (+) US/TiO_2. Toluene concentration: 0.60 ppm; gas flow rate: 1.0 L min^{-1}; TiO_2 amount: 1.0 g L^{-1}.

The trapping efficiency for the WSOI of toluene in the effluent gas by means of the AW (Figure 12) indicated that the quantity of organic intermediates collected in the water was apparently reduced after passing through the AW, implying that they were trapped in the AW. These results clearly indicate that UV pre-treatment for water-insoluble pollutants effectively degraded them and converted them to water-soluble materials that could be easily removed by the AW.

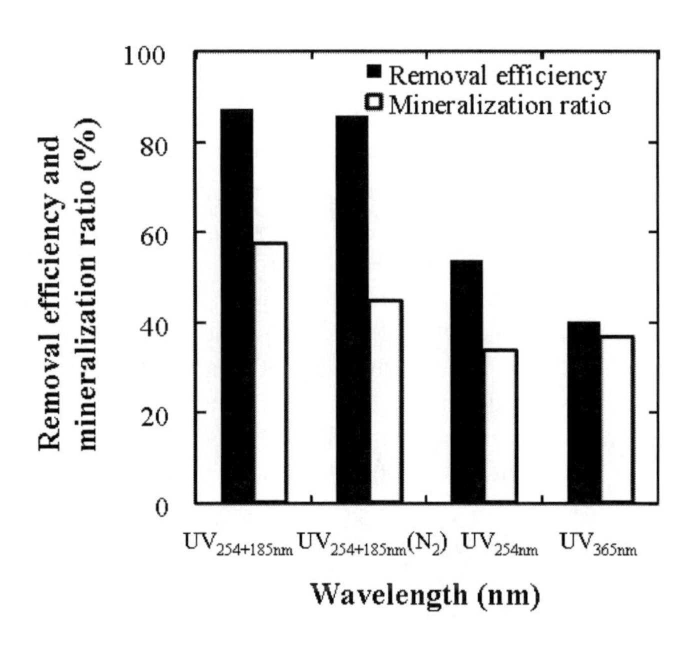

Figure 15. Degradation of toluene under US/UV/TiO_2 conditions for different wavelengths of UV lamps. Toluene concentration: 4.0 ppm; gas flow rate: 1.0 L min^{-1}; TiO_2 concentration: 1.0 g L^{-1}.

3.2.4. Evaluation of TiO_2 Activity under $UV_{254+185nm}$ Irradiation

To enhance the photodegradation of NOx and toluene, we used TiO_2 in combination with $UV_{254+185nm}$ irradiation. Using TiO_2 enhanced the removal of NO_2 and toluene under all reaction conditions (Figure 13).

We supposed that the enhancement was responsible for the photocatalytic oxidation power of TiO_2. In our previous studies, we have reported some of the advantages of combining TiO_2 and $UV_{254+185nm}$ irradiation, such as increased water solubility of organic intermediates, prevention of catalyst deactivation, and inhibited formation of undesirable by-products [22,42] which shows tendencies with the results obtained in this Section.

3.3. Degradation of Organic Gases Using Ultrasonic Mist Generated from TiO_2 Suspension

3.3.1. Toluene Gas Removal

1) Photocatalytic Degradation

Time profiles of toluene removal efficiencies under four reaction conditions, namely, UM alone (US), US/TiO_2, UV_{365nm}/TiO_2 and US/UV_{365nm}/TiO_2, are shown in Figure 14. Toluene gas was not removed under the US or US/TiO_2 conditions since toluene was not captured in the UM or UMP under these conditions due to its hydrophobicity. When toluene gas and the TiO_2 suspension were irradiated with UV_{365nm} (UV_{365nm}/TiO_2 condition), some of the toluene was degraded by photocatalytic reaction at the gas–liquid interface. This result indicates that TiO_2 particles in the liquid phase were sufficiently dispersed onto the gas–liquid interface, and that reactive species were formed by photocatalysis with TiO_2, such as OH radicals, which were present not only at the TiO_2 surface but also in the gas phase close to the liquid phase.

The emission of such reactive species in the gas phase from the TiO_2 surface has been reported previously [43,44]. In addition, under the US/UV/TiO_2 condition, the TiO_2 particles in UMP were also found to be dispersed onto the gas–liquid interface of water droplets, as inferred from the effective degradation of toluene by photocatalysis on the UMP surface. Therefore, water-soluble decomposition intermediates could be generated from toluene by the photocatalytic reaction on the mist surface. However, these intermediates were completely captured in the UMP under this experimental condition, since they were not detected at the reactor exit during the experimental period. From these results, it was thought that the intermediates were continuously decomposed and mineralized to CO and CO_2 in the UMP or the TiO_2 suspension. From this result, it was confirmed that the reaction was steady enough 180 min after it had been started. Based on this, we measured the removal efficiency and mineralization ratio after 180 min passed for further experiment in this study.

2) Effects of UV Wavelength

The removal efficiency and mineralization ratio of toluene for three types of UV lamps with differing wavelengths (UV_{365nm}, UV_{254nm} and $UV_{254+185nm}$) are shown in Figure 15.

The removal efficiency improved with decreasing wavelength of the UV lamp, and the highest removal efficiency and mineralization ratio were obtained under $UV_{254+185nm}$ irradiation. The actual mineralization ratio was expected to be higher than the observed mineralization ratio because CO was hardly detected in this study, and a portion of CO_2 could be easily taken up by the mist as HCO_3^-. The different removal efficiencies obtained under UV_{254nm} and UV_{365nm} irradiation were attributable to the different spectra of the light sources. UV_{365nm} irradiation has a continuous spectrum, and thus light is emitted at wavelengths that cannot be utilized in the photocatalytic reaction. In contrast, UV_{254nm} irradiation has a bright line spectrum that is more suitable for generating OH radicals than UV_{365nm} irradiation. Under $UV_{254+185nm}$ irradiation with 3% of the output at 185 nm, the removal efficiency and mineralization ratio markedly improved, even though the major wavelength was the same as that of UV_{254nm}. This result was interpreted as follows. When the TiO_2 photocatalyst was irradiated by light at wavelengths less than 390 nm, the photocatalytic reaction occurred; thus, the fundamental photocatalytic reactivity of UV_{254nm} and $UV_{254+185nm}$ irradiation was the same. In the case of $UV_{254+185nm}$ irradiation, however, other gas phase reactions occurred. The other gas phase reactions possibly included the generation of active species such as O_3, O and OH radical [19,22], as shown in Eqs. (3)–(7) in Section 3.1.1 and in Eqs. (10) in this Section; these active species are directly involved in the photolysis of organic gases.

$$H_2O + h\nu\ (175 - 190\ \text{nm}) \rightarrow H + \cdot OH \qquad (10)$$

To clarify the contribution of O_3 and O radical, the removal efficiency and mineralization ratio of toluene were measured under a nitrogen atmosphere without O_2. This result is shown in Figure 15. The removal efficiency under the N_2 atmosphere was similar to that in air (N_2:O_2 = 4:1); thus, the contribution of O_3 and O radical to the removal efficiency was low. On the other hand, the mineralization ratio under the N_2 atmosphere without O_2 decreased about 13%.

Direct photolysis and abundant OH radical generation via the reaction shown in Eq. (10) were thought to contribute considerably to the removal efficiency, while O_3 and O radical were thought to contribute to the mineral ization of the degradation products. These results indicate that $UV_{254+185nm}$ irradiation, which could promote both production of OH radical and direct photolysis, was effective in this reactor.

3) Trapping of Decomposition Intermediates

As discussed above, gaseous intermediate products could be trapped in UMP. However, intermediate degradation products are known to be produced both as gases and particulates under $UV_{254+185nm}$ irradiation [22]. The distribution of number concentration of the generated particles at the reactor outlet was measured for UMP (US/UV/TiO_2) or without UMP (UV/TiO_2). The results are shown in Figure 16. A large number of fine particles, which had a peak of about 80 nm, were observed without UMP. Intermediate degradation products such as dicarboxylic acid, which was generated by an oxidative decomposition process of toluene gas, condensed and formed secondary particles [23]. On the other hand, the number concentration of these particles was reduced to ambient levels with UMP.

Most of the particle generation would not occur since soluble precursor gases would be captured in the mist before secondary particles were generated, although a portion of the generated particles might have been trapped by the cooler condenser for UMP at the reactor outlet. It was found that gaseous and particulate intermediate degradation products could be suppressed, even when short-wavelength UV light was used for toluene degradation.

3.3.2. Effect of Various Organic Gases on Removal and Mineralization

Degradation of gaseous organic pollutants by UMP was performed with a specific system by water mist containing TiO_2 photocatalyst. In order to facilitate the development of the proposed system for the degradation of organic gases, we next examined the effectiveness of UMP with TiO_2 photocatalyst at degrading various organic gases. In this study, the degradation of toluene, styrene, *p*-xylene, formaldehyde and acetaldehyde gases was compared under US/UV/TiO_2 conditions and their properties can be described as follows: octanol–water partition coefficient (Log P_{OW}); 2.7, 3.2, 3.2, 0.35 and 0.43, reaction rate constant with OH radical ($k_{OH} \times 10^{-12}$ cm^3 $molecule^{-1}$ s^{-1}); 8.5, 100, 12, 0.20 and 16, and reaction rate constant with O_3 ($k_{O3} \times 10^{-19}$ cm^3 $molecule^{-1}$ s^{-1}); 0.12, 170, 0.0040, 0.000021 and 0.060, respectively. The results of removal experiments using various UV lamps (UV_{365nm}, UV_{254nm} and $UV_{254+185nm}$) are shown in Figure 17. The removal efficiency improved as the wavelength of the UV lamp decreased for all the organic gases examined in this study. In addition, the highest removal efficiency was obtained under

$UV_{254+185nm}$ irradiation. It was thought that the production of OH radical and direct photolysis was promoted by $UV_{254+185nm}$ irradiation [22]. Among the hydrophobic gases (styrene, toluene, *p*-xylene), the highest removal efficiency was observed for styrene, which was the most hydrophobic. This indicates that this degradation process can be applied regardless of hydrophobicity. Furthermore, the rank order of removal efficiency for the hydrophobic gases was toluene < *p*-xylene < styrene (Figure 15) demonstrating that the removal efficiency of the VOC gases depended on their reaction rate with OH radical, without being affected by existing O_3 and O radical as shown in Figure 15. Therefore, the main reaction pathway of VOC degradation was due to OH radical. Formaldehyde was found to have the highest removal efficiency for all the experimental gases, even though it has the slowest reaction rate with OH radical.

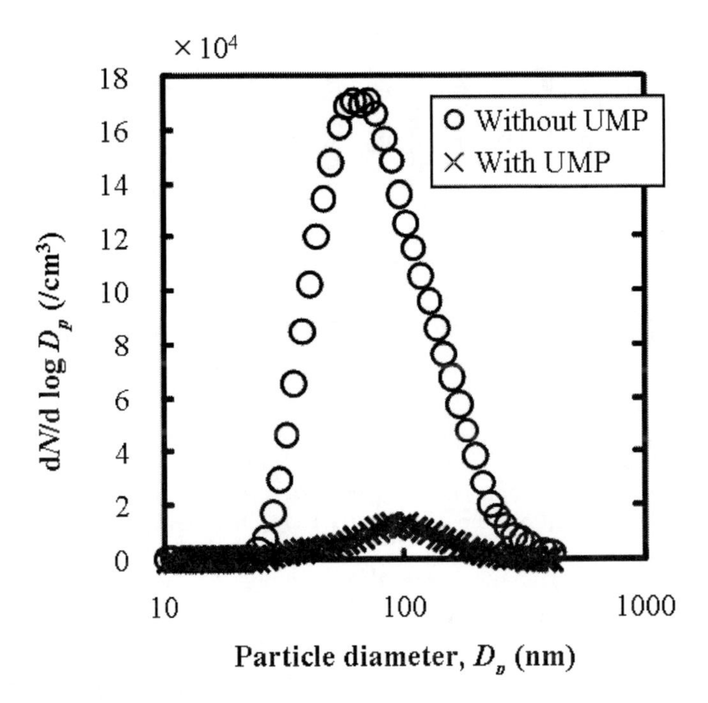

Figure 16. Number concentration (*N*) as function of particle diameter (*D*p) of secondary particles formed from toluene decomposition intermediates under $UV_{254+185nm}$ irradiation. Toluene concentration: 4.0 ppm; gas flow rate: 1.0 L min^{-1}; TiO_2 concentration: 1.0 g L^{-1}.

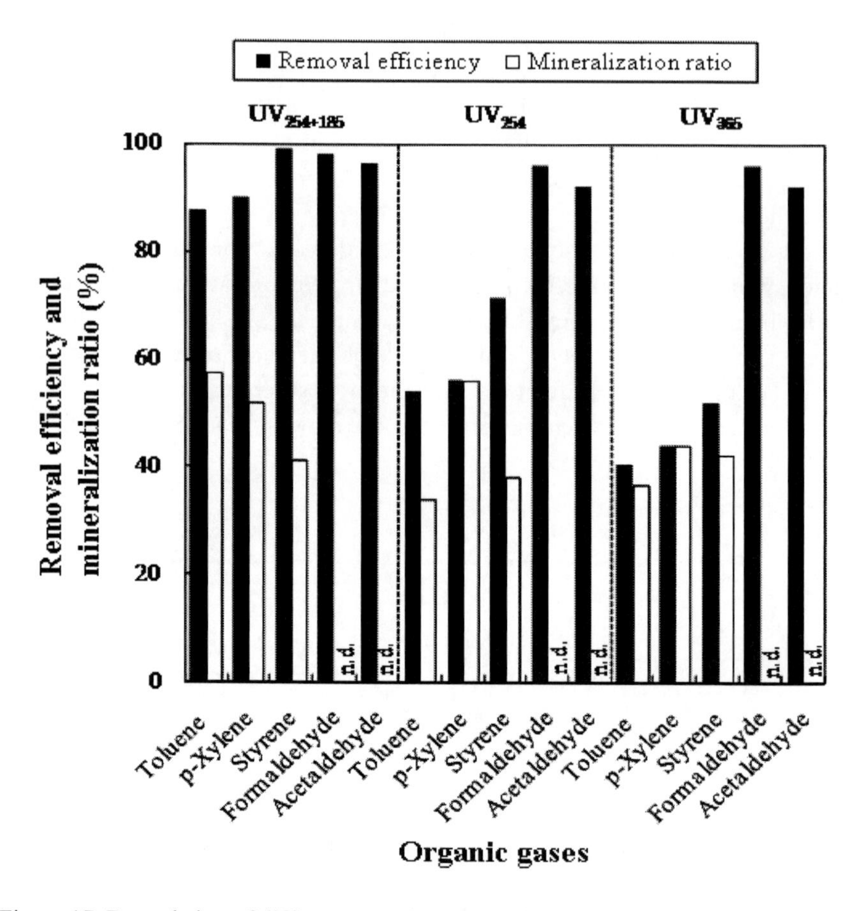

Figure 17. Degradation of different organic gases under US/UV/TiO_2 conditions using UV lamps with different wavelengths. [n.d.: Not detected. Formaldehyde and acetaldehyde gases were immediately absorbed into the water mist.] Initial concentration: 4.0 ppm; flow rate: 1.0 L min^{-1}, TiO_2 concentration: 1.0 g L^{-1}.

High removal efficiency was also obtained for acetaldehyde with the reaction rate nearly the same as *p*-xylene. This shows that acetaldehyde has a similar removal process as formaldehyde by the UMP. It was believed that formaldehyde and acetaldehyde were taken up by the mist faster than they were degraded, since formaldehyde and acetaldehyde are highly water soluble. Therefore, the mineralization was not detected in gas phase due to insufficient CO and CO_2, although the removal efficiency of formaldehyde and acetaldehyde were high. These results indicate that highly water-soluble gases such as formaldehyde and acetaldehyde are degraded after being taken up by the mist and subsequently undergone mineralization in the liquid phase.

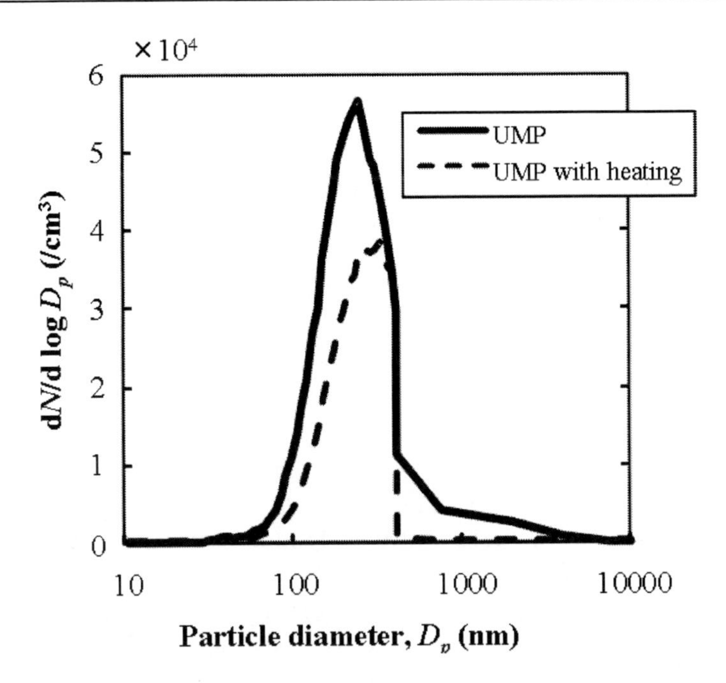

Figure 18. Number concentration (N) as a function of particle diameter (Dp) for UMP with or without heating by an electric furnace. Gas flow rate: 1.0 L min^{-1}; TiO_2 concentration: 1.0 g L^{-1}.

3.3.3. Size Distribution of UMP

The size distributions of UMP without or with heating in the electric furnace of 200°C for about 10 s are shown in Figure 18.

It was found that the water droplets in UMP had particle diameters on the submicron to micron scale, and the peak number concentration was observed around 200–300 nm. When UMP was heated until water was completely removed, the particle size distribution decreased to the submicron scale and was found out to be similar to the UMP peak diameter even though a portion of ultrafine particles with the diameter below 100 nm might have been lost by the diffusion adhesion. It was reported in our previous study that TiO_2 particles are also incorporated into micron-scale water droplet [25]. It was thought that the UMP was a mixture of water droplets on the submicron and micron scale and that these droplets included TiO_2 particles coagulated on the submicron scale. Therefore, the submicron-scale UMP like made the dominant contribution to the photocatalytic reaction, based on the number concentration of UMP and the contact with TiO_2 on the surface. For the past several decades, it has been widely believed that liquid droplets generated by ultrasonic atomization have diameters on the order of μm [45,46]. However, it was found that submicron-scale UMP makes the dominant contribution to the photocatalytic reaction examined in this study, and in the recent study, which was related to mist production and real-time measurement of mist size distribution, the remarkable observation of submicron-scale mist production was also reported [47].

3.3.4. Degradation Mechanism for Organic Gases on UMP Surface

The mechanism of the degradation process for gaseous organic pollutants with UMP by the US/UV/TiO_2 method is proposed as shown in Figure 19. The degradation reaction under US/UV/TiO_2 conditions primarily occurred on the surface of submicron-scale water droplets, while the mist was a mixture of micron- and submicron-scale water droplets.

The mist included TiO_2 that coagulated to form submicron-scale particles, even after drying. These TiO_2 particles participated in the photocatalytic reaction on the surface of UMP droplets. Hydrophobic and hydrophilic organic gases were effectively degraded via oxidation by the active mist at the water droplet surface or inside of the water droplet after dissolution. The water-soluble degradation intermediates generated from hydrophobic organic gas were also captured by the water droplets, and the captured degradation intermediates were completely decomposed in the liquid phase with continuous recovery of the TiO_2 activity. Based on these degradation mechanisms, high-efficiency treatment of organic gases could be realized completely in the gas phase.

CONCLUSIONS

The photodegradation of toluene and benzene with TiO_2 and $UV_{254+185nm}$ irradiation from an ozone-producing UV lamp was introduced as a first step to show VOC degradation in gas phase using short-wavelength UV irradiation with TiO_2 catalyst with different UV sources.

The feasibility of using short-wavelength UV irradiation with TiO_2 catalyst in photodegradation of gaseous toluene was evaluated by comparing three UV sources (UV_{365nm}, UV_{254nm}, and $UV_{254+185nm}$). Our main aim was to determine if some crucial disadvantages associated with classical photocatalytic degradation could be avoided by using short-wavelength UV irradiation.

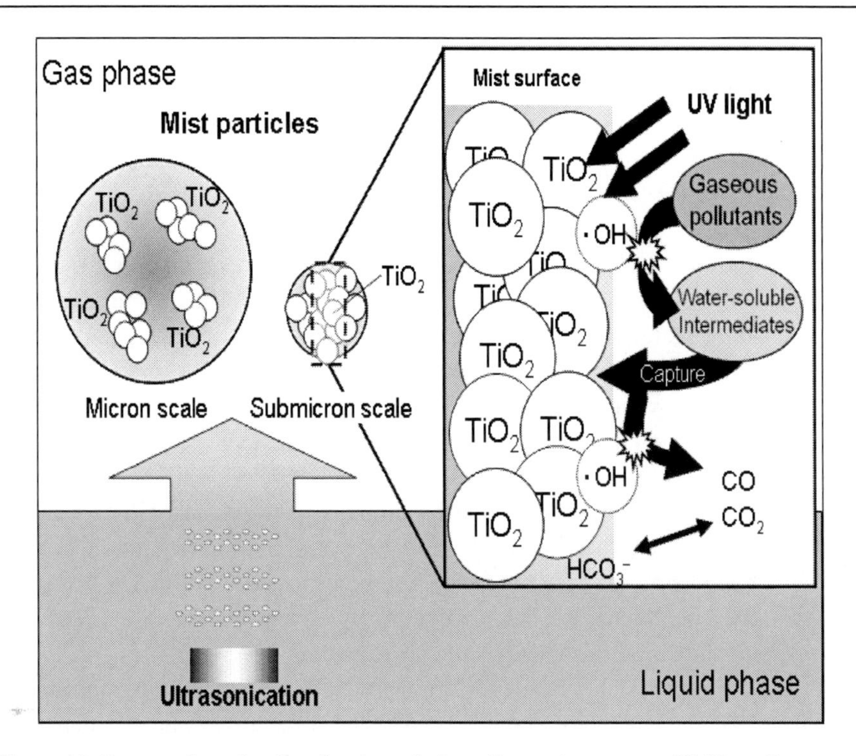

Figure 19. Proposed mechanism for degradation of organic gases on UMP surface.

Our main conclusions are as follows:

1. The use of TiO_2 under $UV_{254+185nm}$ irradiation significantly enhanced the photodegradation of VOCs relative to UV alone, owed to the combined effect of photochemical oxidation in the gas phase and photocatalytic oxidation on the TiO_2 surface.
2. The highest conversion of toluene and mineralization was obtained over TiO_2 catalyst with $UV_{254+185nm}$ irradiation among the tested three UV sources.
3. The use of $UV_{254+185nm}$ irradiation effectively prevented deactivation of the TiO_2 catalyst, even at high toluene concentrations, which has been a substantial obstacle for the industrial application of classical photocatalytic degradation.
4. Reactive oxygen species (O_2^-, $O(^1D)$, and O_3) appeared to play an decisive role with OH radicals in the complete oxidation of less-reactive intermediates over the TiO_2 catalyst.

5. Photogenerated excess ozone and residual organic compounds were simultaneously eliminated by using an ozone-decomposition catalyst (ODC), suggesting that an ODC reactor can be coupled with a photocatalytic reactor employing $UV_{254+185nm}$ irradiation for air purification.

This study revealed that the products from the photodegradation of VOCs with the $UV_{254+185nm}$ photoirradiated TiO_2 were mainly mineralized CO_2 and CO, but some water-soluble organic intermediates were also formed under more severe reaction conditions. The water-soluble aldehydes and carboxylic intermediates disappeared from the effluent gas stream and were detected in the water impingers suggesting that the intermediates can be washed out by conventional gas washing technique, such as wet scrubber or air washer.

As a second step, we introduced removal of water-insoluble gaseous pollutants (NOx and toluene) with $UV_{254+185nm}$ irradiation in humid air to provide a useful process for effectively removing gaseous pollutants from the air and the treatment of intermediates by trapping the water-soluble intermediates into water droplets by the air washer.

A useful process for the degradation of NOx and toluene by means of $UV_{254+185nm}/TiO_2$ irradiation coupled with the use of an air washer was presented. The OH radicals and ozone produced by $UV_{254+185nm}$ irradiation effectively degraded NOx and toluene to HNO_3 and CO_2, respectively. The organic intermediates formed during toluene degradation were highly water soluble and could therefore be effectively removed, along with the HNO_3, by the air washer. The use of TiO_2 along with the $UV_{254+185nm}$ irradiation increased the photodegradation of pollutants by means of photocatalytic oxidation. The combination of short wavelength UV/TiO_2 treatment and an air washer is a useful process for effectively removing gaseous pollutants from the air, particularly those with high reactivity with OH radicals and ozone. However, using the air washer as a means for effective removal of gaseous pollutants and their intermediates has disadvantages for example, 1) the reaction can be completed with 2-step process and 2) the size of air washer is relatively large. For these reasons, finally we studied and described VOCs degradation using an ultrasonic mist generated from TiO_2 suspension under UV_{365nm}, UV_{254nm}, and $UV_{254+185nm}$ irradiations. With this technology, gaseous pollutants and the intermediates could be degraded and captured by water droplet and decomposed further in a liquid phase with 1-step process by generating mist droplets from ultrasonic atomization of TiO_2 suspension.

Toluene, which has poor water solubility, reacted with TiO_2 present on the surface of the water droplets, and high removal efficiency was obtained by using UV irradiation and UMP. Water soluble intermediate products could be taken up by water droplets, and as a result, intermediate degradation products were not observed at the reactor exit.

The removal efficiency of gaseous organic pollutants by UMP was improved by using UV light with shorter wavelengths. In particular, when $UV_{254+185\ nm}$ irradiation was used, not only removal but also mineralization, whereby organic gases are converted to CO and CO_2, was improved. In this case, O_3 and O radical more effectively reacted to bring about mineralization of the intermediate degradation products compared with the removal of the introduced gaseous organic pollutant. When applying the UMP method to organic gases having different properties, hydrophobic gas could be decomposed regardless of hydrophobicity, even though the reaction occurred on the surface of water droplets. The sufficient removal efficiency obtained when using UMP was related to the reaction rate with OH radical. On the other hand, highly water-soluble organic gases were taken up by the mist faster than the degradation reaction occurred, and were effectively removed by the mist.

We proposed a mechanism for the degradation of organic gas using UMP. The mist contained a mixture of micron- and submicron-scale water droplets, and the main reaction was caused by submicron-scale water droplet including sufficient TiO_2 particles and organic gas for effective oxidative decomposition on the surface or inside of UMP. The water-soluble degradation products generated from the hydrophobic gas could also be captured by the water droplets. Therefore, effective and complete treatment of pollutant gases could be achieved.

Based on these results, an air purification technique utilizing UMP has the potential to be effective for not only simultaneous removal of gaseous organic pollutants with hydrophobic and/or hydrophilic property, but also suppression of degradation products in the exhaust. Therefore, this method is thought to have high potential as an air purification technique.

REFERENCES

[1] Alberici, R.M., Jardim, W.F., 1997. *Appl. Catal. B Environ.* 14, 55–68.

[2] Hennezel, O., Ollis, D.F., 1997. *J. Catal.* 167, 118–126.

[3] Ao, C.H., Lee, S.C., Mak, C.L., Chan, L.Y., 2002. *Appl. Catal. B Environ.* 42, 119–129.

[4] Ameen, M.M., Raupp, G.B., 1999. *J. Catal.* 184, 112–122.

[5] Martra, G., Coluccia, S., Marchese, L., Augugliaro, V., Loddo, V., Palmisano, L., Schiavello, M., 1999. *Catal. Today* 53, 695–702.

[6] Cao, L., Gao, Z., Suib, S.L., Obee, T.N., Hay, S.O., Freihaut, J.D., 2000. *J. Catal.* 196, 253–261.

[7] Blount, M.C., Falconer, J.L., 2002. *Appl. Catal. B Environ.* 39, 39–50.

[8] Einaga, H., Futamura, S., Ibusuki, T., 2002. *Appl. Catal. B Environ.* 38, 215–225.

[9] Ray, M.B., 2000. *Dev. Chem. Eng. Miner. Process.* 8, 405–439.

[10] Maira, A.J., Yeung, K.L., Lee, C.Y., Yue, P.L., Chan, C.K., 2000. *J. Catal.* 192, 185–196.

[11] Li, X.Z., Li, F.B., Yang, C.L., Ge, W.K., 2001. *J. Photochem. Photobiol. A Chem.* 141, 209–217.

[12] Muggli, D.S., Ding, L., 2001. *Appl. Catal. B Environ.* 32, 181–194.

[13] Ichiura, H., Kitaoka, T., Tanaka, H., 2003. *Chemosphere* 51, 855–860.

[14] Zhao, J., Yang, X., 2003. *Build. Environ.* 38, 645–654.

[15] Mazzarino, I., Piccinini, P., Spinelli, L., 1999. *Catal. Today* 48, 315–321.

[16] Wang, S., Shiraishi, F., Nakano, K., 2002. *Chem. Eng. J.* 87, 261–271.

[17] Kohno, H., Tamura, M., Sakamoto, K., 1996b. *J. Aerosol Res. Jpn.* 11, 349–356 (in Japanese).

[18] Sekiguchi, K., Ishitani, O., Sakamoto, K., 1999. *In: Proceeding of the 8th International Conference on Indoor Air Quality Climate*. Edinburgh, Scotland, pp. 695–700.

[19] Feiyan, C., Pehkonen, S.O., Ray, M.B., 2002. *Water Res.* 36, 4203–4214.

[20] Wang, J.H., Ray, M.B., 2000. *Sep. Purif. Technol.* 19, 11–20.

[21] Kohno, H., Tamura, M., Sakamoto, K., 1996a. *J. Jpn. Air Clean. Assoc.* 34, 37–44 (in Japanese).

[22] Sekiguchi, K., Yamamoto, K., Sakamoto, K., 2008. *Catal. Commun.* 9, 281–285.

[23] Jeong, J., Sekiguchi, K., Sakamoto, K., 2004. *Chemosphere* 57, 663–671.

[24] Jeong, J., Sekiguchi, K., Lee, W., Sakamoto, K., 2005. *J. Photochem. Photobiol. A* 169, 279–287.

[25] Jeong, J., Sekiguchi, K., Saito, M., Lee, Y., Kim, Y., Sakamoto, K., 2006. *Chem. Eng. J.* 118, 127–130.

[26] Williams II, E.L., Grosjean, D., 1990. *Environ. Sci. Technol.* 24, 811–814.

[27] Hirakawa, T., Yawata, K., Nosaka, Y., 2007. *Appl. Catal., A* 325, 105–111.

[28] Ohno, T., Tokieda, K., Higashida, S., Matsumura, M., 2003. *Appl. Catal., A* 244, 383–391.

[29] Bowering, N., Walker, G.S., Harrison, P.G., 2006. *Appl. Catal. B* 62, 208–216.

[30] Sekiguchi, K., Morinaga, W., Sakamoto, K., Tamura, H., Yasui, F., Mehrjouei, M., Muller, S., Moller, D., 2010. *Appl. Catal., B* 97, 190–197.

[31] Sakamoto, K., Tonegawa, Y., Ishitani, O., 1999a. *J. Adv. Oxid. Technol.* 4, 35–39.

[32] Sakamoto, K., Fukumuro, T., Ishitani, O., Kohno, H., 1999b. *In: Proceeding of the 8th International Conference on Indoor Air Quality Climate*. Edinburgh, Scotland, pp. 673–678.

[33] Wang, K.H., Tsai, H.H., Hsieh, Y.H., 1998. *Chemosphere* 36, 2763–2773.

[34] Jo, W.K., Park, J.H., Chun, H.D., 2002. *J. Photobiol. A Chem.* 148, 109–119.

[35] Villasenor, J., Reyes, P., Pecchi, G., 2002. *Catal. Today* 76, 121–131.

[36] Sekiguchi, K., Sanada, A., Sakamoto, K., 2003. *Catal. Commun.* 4, 247–252.

[37] Futamura, S., Zhang, A., Einaga, H., Kabashima, H., 2002. *Catal. Today* 72, 259–265.

[38] Hao, Z., Cheng, D., Guo, Y., Liang, Y., 2001. *Appl. Catal. B Environ.* 33, 217–222.

[39] Calvert, J.G., Atkinson, R., Becker, K.H., Kamens, R.M., Seinfeld, J.H., Wallington, T.J., Yarwood, G., 2002. Oxford University Press, Oxford.

[40] Sekiguchi, K., Sakamoto, K., Fujii, T., 1998. *J. Aerosol Res. Jpn.* 13, 222–229.

[41] Peukert, W., Wadenpohl, C., 2001. *Powder Technol.* 118, 136–148.

[42] Jeong, J., Sekiguchi, K., Sakamoto, K., 2004. *J. Photochem. Photobiol. A: Chem.* 169, 277–285.

[43] Tatsuma, T., Tachibana, S., Miwa, T., Tryk, D.A., Fujishima, A., 1999. *J. Phys. Chem. B* 103, 8033–8035.

[44] Tatsuma, T., Tachibana, S., Fujishima, A., 2001. *J. Phys. Chem. B* 105, 6987–6992.

[45] Rajan, R., Pandit, A.B., 2001. *Ultrasonics* 39, 235–255.

[46] Yasuda, K., Bando, Y., Yamaguchi, S., Nakamura, M., Oda, A., Kawase, Y., 2005. *Ultrason. Sonochem.* 12, 37–41.

[47] Kobara, H., Tamiya, M., Wakisaka, A., Fukazu, T., Matsuura, K., 2009. *AIChE J.* 56, 810–814.

In: Volatile Organic Compounds
Editors: J. C. Hanks et al. pp. 89-118
ISBN 978-1-61324-156-1

Chapter 3

CATALYTIC INCINERATION OF VOLATILE ORGANIC COMPOUNDS

Ting Ke Tseng, Yi Hsing Lin and Hsin Chu*
Department of Environmental Engineering and Sustainable Environmental Research Center, National Cheng Kung University, Tainan 701, Taiwan

ABSTRACT

Volatile organic compounds (VOCs) make up a major class of air pollutants. This class includes pure hydrocarbons, partially oxidized hydrocarbons (organic acids, aldehydes, ketone), as well as organics containing chlorine, sulfur, nitrogen, or other elements. These compounds are usually found in most manufacturing processes, either for the raw materials, intermediates, or the finished products. Organic materials are present as chemicals, solvents, release agents, coatings, decomposition products, pigments, and so on that eventually must be disposed.

Catalytic incineration is a well known process to destruct VOC emissions in air at low energy cost; it is useful when these contaminants are toxic and malodorous. The advantages of catalytic oxidation can be important mainly because of potential savings following the lower temperatures required. The environmental impact can be improved because both higher efficiency of abatement and lower levels of NO_x and

* To whom all correspondence should be addressed. Tel: (886)–6–208 0108; Fax: (886)–6–275 2790; E–mail: chuhsin@mail.ncku.edu.tw

CO_2 emissions can be reached. Furthermore the small pressure drops make the process very attractive.

This article aims to show and discuss the results from continuous tests conducted in the laboratory scale reactors. Measurements of the abatement in the simulated industrial conditions, the assessment of the catalyst aging, and the identification of the possible poisoning will be shown.

The catalyst powders were prepared by the incipient wetness impregnation method with aqueous solutions of metal nitrate and calcined at proper temperatures. The finished catalysts were characterized first by DTA–TGA.

The effects of operating parameters, such as inlet temperature, space velocity, VOCs inlet concentration, and oxygen concentration on the catalytic incineration of VOCs over the catalysts were then performed.

The activity of the catalyst decreased significantly with time while VOCs incineration was operated under a low temperature. However, the activity of the catalyst did not change much while the operating temperature was high. The catalysts were characterized by the surface and pore size analysis, XRD, XPS, EDS and SEM before and after the tests.

Three kinetic models (i.e., the power–rate law, the Mars and Van Krevelen model, and the Langmuir–Hinshelwood model) were used to analyze the results. A differential reactor design was used for best fit of kinetic models in this study.

1. INTRODUCTION

Volatile organic compounds (VOCs) make up a major class of air pollutants. This class includes pure hydrocarbons, partially oxidized hydro carbons (organic acids, aldehydes, ketone), as well as organics containing chlorine, sulfur, nitrogen, or other elements. VOCs are defined as organic compounds that have high vapor pressure and easily vaporize at the condition of ambient temperature and pressure. The VOCs are a wide–ranging class of chemicals derived from many sources and contain over 300 hazardous compounds as designated by the United States Environmental Protection Agency (Taylor et al., 2000). These compounds are usually found in most manufacturing processes, either for the raw materials, intermediates, or the finished products. Organic materials are present as chemicals, solvents, release agents, coatings, decomposition products, pigments, and so on that eventually must be disposed.

In such manufacturing processes, there is usually a gaseous effluent that contains low concentrations of organics and is vented into the atmosphere. Examples of commercial operations that produce VOCs emissions are

- Chemical plants
- Pharmaceutical plants
- Airplane manufacturers
- Fiber manufacturers
- Printing plants
- Wire enameling plants
- Painting facilities
- Petroleum refineries
- Automobile manufacturers
- food processors
- textiles manufacturers
- can coating plants
- electronic component plants
- wood stoves

Besides causing harmful effects on human organs, VOCs may also react with NOx in the atmosphere to form even more toxic photochemical smog and O_3 (de Nevers, 2000).

Some VOCs, such as mercaptans (methyl mercaptan, ethyl mercaptan), dimethyl sulfides, dimethyl disulfide, styrene, acrylonitrile, methyl isobutyl ketone and halogenated volatile organic compounds are not only extremely toxic but also may give off an offensive odor even at an extremely low concentration.

The increasing stringent environmental regulations limiting emissions of halogenated hydrocarbons have highly increased the demand for technologies to remove such halogenated compounds efficiently from waste streams. The current technology for the complete oxidation of chlorinated hydrocarbons is thermal incineration, which requires extremely high temperatures of about 1000°C (Artizzu–Duart et al., 1999). While it is a simple and often effective method of control, the high temperatures required culminate in a relatively fuel intensive technique with little control over the ultimate products. The latter is particularly problematic and can result in incomplete oxidation of the waste stream and the formation of toxic by–products such as dioxins, dibenzofurans and oxides of nitrogen, if conditions are not carefully controlled.

Some people believe that the best technique to eliminate these toxic materials in the waste stream is catalytic incineration (Toledo et al., 2001). Heterogeneous catalytic incineration has received the most attention lately because it is a final disposal and energy saving process (van der Vaart et al., 1991a, van der Vaart et al., 1991b). However, chlorinated or sulfur-containing VOCs may deactivate the catalyst and reduce the advantage of catalytic incineration.

Catalysts reported for destructive oxidation of chlorinated VOCs mostly consist of base metal oxides and noble metals on acidic supports (Agarwal et al., 1992, Bickle et al., 1994, Rossin and Farris, 1993). Most research has focused on noble metal catalysts and very little has been reported on combustion of chlorinated VOC over supported base metal oxides.

The supported noble metal systems show high activity for the oxidation of many VOCs, with high selectivity to carbon oxide products. However, these tend to be relatively expensive and can be rapidly deactivated by the presence of chlorinated compounds, sulfur or other metals in the waste stream. The second class of catalysts are metal oxides and some of the most active ones are based on copper (Kang and Wan, 1994), cobalt (Drago et al., 1995), chromium (Agarwal et al., 1992) and manganese (Watanabe et al., 1996).

Although the poisoning effect of chlorine on the catalysts has been studied thoroughly, the poisoning effect of sulfur on the catalysts is also interesting and should be explored. The scope of the present work is summarized in Figure 1. In addition, the main objectives of the present work are summarized as follows:

(1) Study the characteristics of the catalysts.
(2) Study the effect of operating parameters and deactivation of the VOCs decomposition on the various catalysts.
(3) Study the best fitting of the kinetic models and the comparison of the predicted and experimental data.

2. THE PREPARATION OF CATALYSTS

Each chemical reaction has a unique *reaction path* due to its unique *activation energy*. At the same temperature, the number of high-energy molecules is usually less; therefore, the reaction frequency decreases, and the reaction rate is slower. A catalyst can change the reaction path to reduce activation energy to accelerate the reaction. As described above, it is very important to understand how to prepare catalysts. The details are described below.

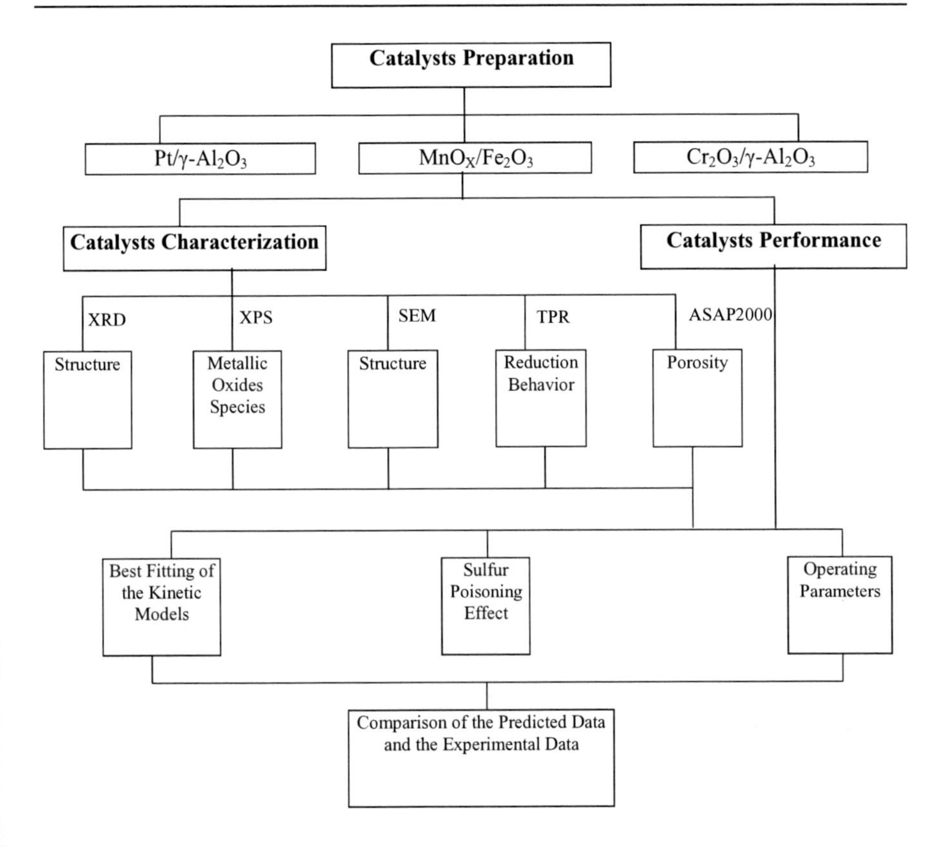

Figure 1. Research scope for catalytic incineration of sulfur-containing VOCs on the MnO_X/ Fe_2O_3 catalysts.

2.1. Catalyst Materials

The composition of a catalyst can be divided into the catalyst material and carrier. According to the metal of catalyst materials, it can be divided into precious metals, transition metals and bi-metal alloys. Precious metals, such as Pt, Pd, Pt, etc., exhibit higher activity than transition metal catalysts. However, they are expensive, easily sintered and poisoned. In recent years, the transition metal-based catalysts have been successfully developed due to their higher thermal stability and lower cost. Most bi-metal catalysts, composed of an active group VIII metals (Fe, Co, Ni, Ru, Rh, Pd, etc.) and an inactive group IB metal (Cu, Ag, Au), have been synthesized as catalysts for a variety of reactions (Chandler et al., 2000).

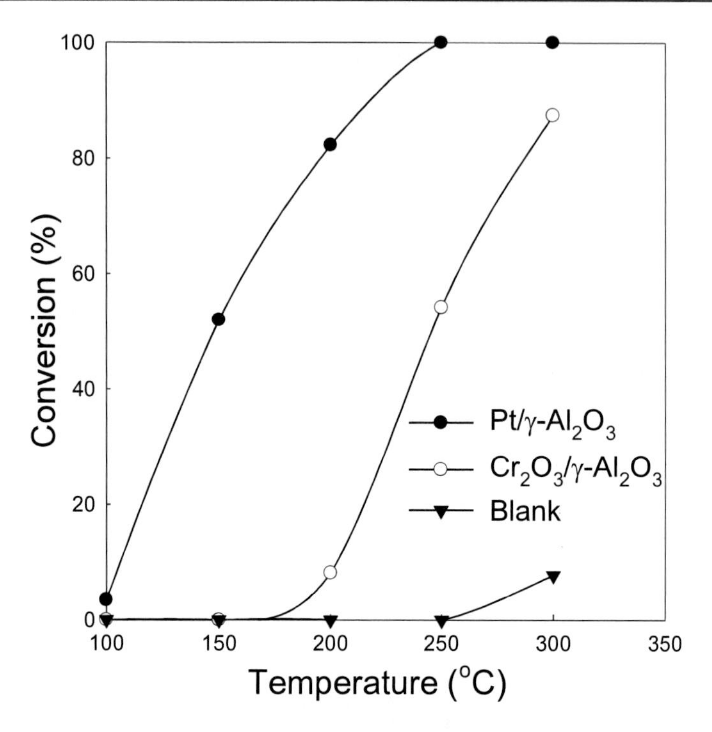

Figure 2. Comparisons of the catalytic incineration of MIBK on the Cr_2O_3/γ-Al_2O_3 catalyst, Pt/γ-Al_2O_3 catalysts and a blank test at various temperatures (inlet concentration: 100 ppm; gas hourly space velocity (GHSV): 80,000 h^{-1}; temperature: 100–300°C; O_2 concentration: 20.8%).

2.2. Carriers

It is well known that most carriers are highly porous materials, and adsorption of VOCs mainly occurs at surface pore or specific surface site. Alumina is one of the most common carriers used in environmental applications. There are many kinds of alumina, having a variety of surface areas, pore size distributions, surface acidic properties, and crystal structures. Its properties depend on its preparation, purity, and thermal history (Satterfield, 1980). The most extensively used carriers are γ-Al_2O_3 and η-Al_2O_3, which have high area and relatively stable thermal conductivity. Similar to Al_2O_3, SiO_2 is also widely used as a carrier which has high surface area that can be as high as 300~400 m^2/g. TiO_2 is an effective selective catalyst to enable the removal of harmful NO_x gases because its surface can absorb light under a wide range of visible light illumination. TiO_2 exists in many crystalline material forms, the most important of which is anatase. The anatase form is the most important because it has the highest surface area (50~80 m^2/g) and is thermally stable up to about 500°C (Kominami et al., 1996). Zeolite is a crystalline alumina-silicate used to form a microporous solid, which has a precise pore structure (Ward, 1967). The small uniform pores and large intracrystalline volumes characteristic of zeolites make them ideally suited as shape-selective catalysts or catalyst carriers.

2.3. Making the Finished Catalyst

The most common preparation procedure for dispersing the catalytic species within the carriers is by impregnating an aqueous solution containing a salt (precursor) of the catalytic element or elements (Thomas and Brundrett, 1980; Stiles, 1983; Komiyama, 1985; Worstell, 1992). According to water content, the method can be divided into incipient wetness and capillary impregnation. The maximum water uptake by the support is referred to as the water pore volume. This is determined by slowly adding water to a carrier until it is saturated, as evident by the beading of the excess H_2O. The active metal precursor is then dissolved in an aqueous or organic solution to the water pore volume. Most preparation methods simply involve soaking the carrier in the solution and allowing capillary and electrostatic forces to disperse the salt over the internal surface of the porous network. The salt generating the cations or anions containing the catalytic element are chosen to be compatible with the surface charge of the carrier to obtain efficient adsorption or, in some cases, ion exchange. The isoelectric point of the carrier (the charge assumed by the carrier surface), which is dependent on pH, is useful in making decisions regarding salts and pH conditions for the preparation.

2.4. Drying

Excess water of physical adsorption and other volatile species can be removed by drying to get some physical properties. The operating temperature is usually at about 110-150°C.

2.5. Calcination and Activation

It is most common to calcine the catalyst in forced air to about 400~500°C to remove all traces of decomposable salts used for preparing the catalyst. The purposes of calcination are usually: (i) it can reduce the water content, and further, to remove the impurities materials (adhesives), volatile materials, and unstable anions; (ii) the temperature of calcination of the catalyst has to achieve the reaction temperature (300-750°C). At this temperature, a sintering phenomenon may occur, which can increase the mechanical strength of the catalyst; and (iii) various materials can be mixed by thermal diffusion during the high temperature calcination process, and the process can urge decomposition of metal salts to further form the crystalline phase. Heat has great influence on the activation of catalysts for the calcination procedure, so if you want to get the most appropriate conditions, you have to constantly try to obtain the temperature.

3. Effect of Operating Parameters

The major parameters affecting catalytic incineration of VOCs include catalyst type, VOC type, VOC concentration, operating temperature, space velocity and O_2 concentration (Spivey, 1987). The following are effects of operating parameters.

3.1. Catalyst Type

The MIBK (methyl-isobutyl-ketone) conversions over the Pt/γ-Al_2O_3 and Cr_2O_3/γ-Al_2O_3 catalysts and a blank test by replacing catalyst packing with glass fiber are shown in Figure 2 (Tseng et al., 2005). The Pt and Cr_2O_3 catalysts are active in the range of 373–573 K. The results indicate that the MIBK conversion over the Pt catalyst is higher than the Cr_2O_3 catalyst at all temperatures in the operating range of this study. The Thermo-gravimetric analyses for the Pt/γ-Al_2O_3 catalyst were recorded. There are significant weight losses from 323 to 373 K, but no large weight loss occurs at higher temperatures no matter what thermo-program was used. The results show that the water adsorbed by the catalyst desorbs from the catalyst at a temperature range of 323–373 K. Above 373 K, the Pt/γ-Al_2O_3 catalyst shows a stable characteristic.

Figure 3 shows comparisons of the catalytic incineration of C_8H_8 on Pt/Al_2O_3, Cr_2O_3/Al_2O_3, and Mn_2O_3/Fe_2O_3 and a blank test at various temperatures. The preliminary tests started from a series of blank tests by replacing catalyst packing with glass fiber and performing a feasibility study on three catalysts. The results show that the conversion of C_8H_8 is trivial in the operating ranges of the blank tests. This suggests that the catalyst is the key element for the conversion of C_8H_8. However, Pt/Al_2O_3 and Cr_2O_3/Al_2O_3 are not as efficient as Mn_2O_3/Fe_2O_3. This is consistent with the results of Chu and Wu (Chu and Wu, 1998) in which they showed MnO/Fe_2O_3 was a better catalyst for conversion of C_2H_5SH.

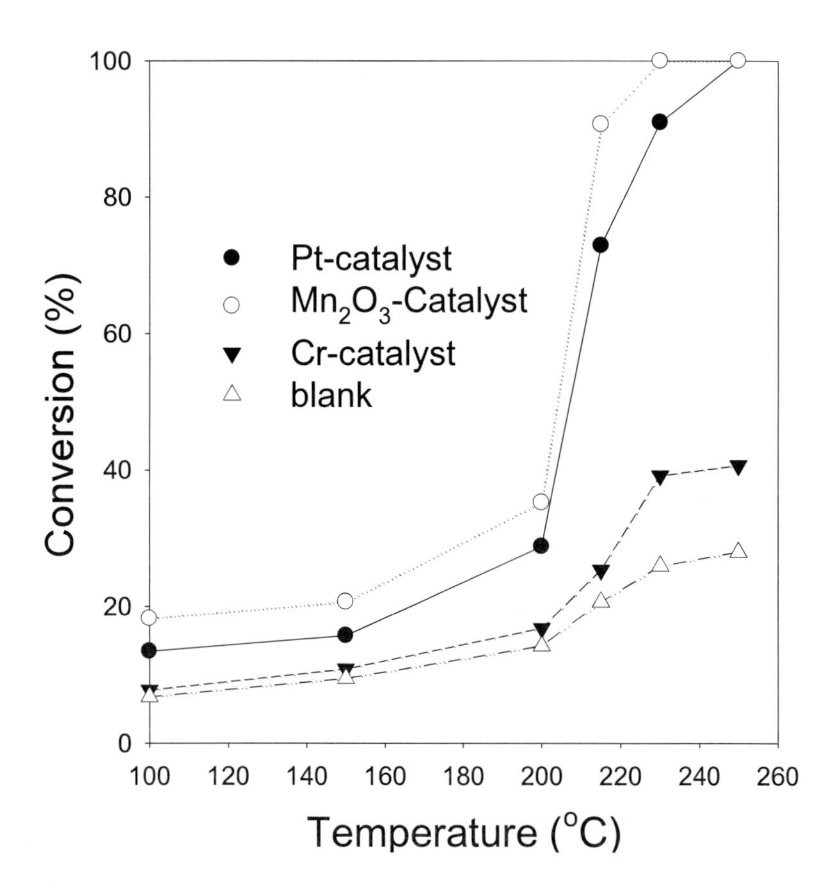

Figure 3. Comparisons of the catalytic incineration of C_8H_8 on $Pt/\gamma\text{-}Al_2O_3$, $Cr_2O_3/\gamma\text{-}Al_2O_3$, and Mn_2O_3/Fe_2O_3 and a blank test at various temperatures (C_8H_8 concentration = 50 ppm; GHSV = 70,000 hr^{-1}; temperature = 100–250°C; O_2 concentration = 20.8%).

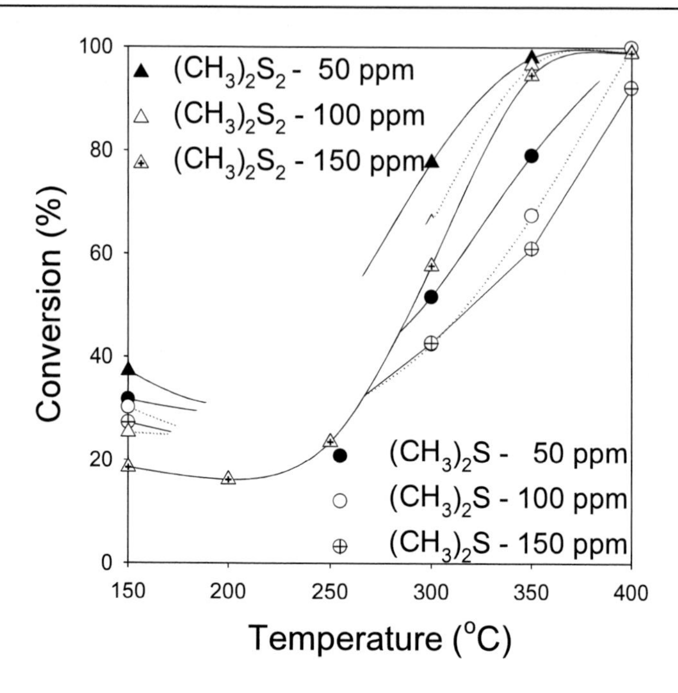

Figure 4. The effect of temperature on the catalytic conversion of $(CH_3)_2S$ and $(CH_3)_2S_2$ over a MnO/Fe_2O_3 catalyst. (space velocity: 55,000 h^{-1}; O_2 concentration: 20.8%).

Heyes et al. (Heyes et al., 1982) evaluated a series of catalysts for the destructive oxidation of n-butanal and methyl mercaptan. They found that the ability to destroy butanal in mixtures with methyl mercaptan at the end of life-tests, decreased in the following order: $CuO = Pt > MnO_2 > V_2O_5 > CO_3O_4$. Völter et al. indicated that the activity of a Pt catalyst was stronger in air than in pure oxygen. These studies revealed that the catalyst type is a key parameter for catalytic incineration (Völter et al., 1987).

3.2. VOCs Type

Destructibility of VOCs could be affected by the property of pollutions and catalysts. Tichenor and Palazzolo did a series of tests on catalytic incineration of VOCs using precious metals. They found that destructibility of VOCs using precious metals decreased in the following order: alcohols > cellusolves > aldehydes > aromatics > ketones > acetates > alkanes > chlorinated hydrocarbons(Tichenor and Palazzolo, 1987).

3.3. Reaction Temperature

The effect of temperature on catalytic activity impacts not only the kinetic reaction, but also the adsorption of contaminants. The effect of temperature on the catalytic conversion of $(CH_3)_2S$ and $(CH_3)_2S_2$ over a MnO/Fe_2O_3 catalyst are shown in Figure 4. This figure shows that the conversion of $(CH_3)_2S$ increases as the inlet temperature increases. The concentration effects are similar to those for $(CH_3)_2S$. However, a higher temperature is required to accomplish the same $(CH_3)_2S$ conversion compared to $(CH_3)_2S_2$. Figure 5 shows the effect of temperature on the catalytic conversion of C_2H_3CN and C_8C_8 over a MnO/Fe_2O_3 catalyst. It suggests that the conversion of C_8H_8 increases as inlet temperature increases in the range of 100–250°C. The same phenomenon is also observed in C_2H_3CN, as the C_2H_3CN inlet concentration 150 ppm, the C_2H_3CN conversion reaches 97% at 250°C and 21% at 100°C. The conversion of C_2H_3CN increases as inlet temperature increases in the range of 100~250°C. These results indicate the conversion efficiency usually increases with increasing temperature.

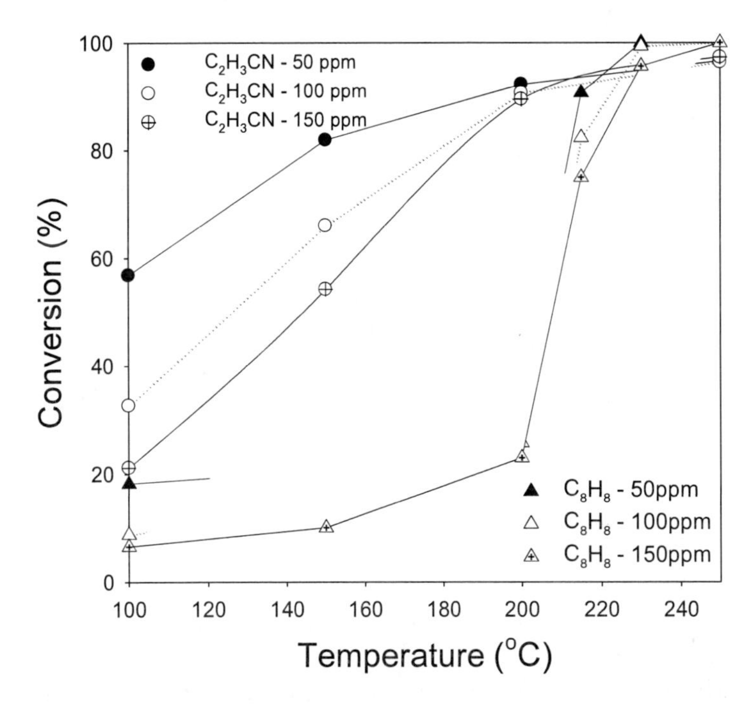

Figure 5. The effect of temperature on the catalytic conversion of C_2H_3CN and C_8C_8 over a MnO/Fe_2O_3 catalyst. (space velocity: 70,000 h^{-1}; O_2 concentration: 20.8%).

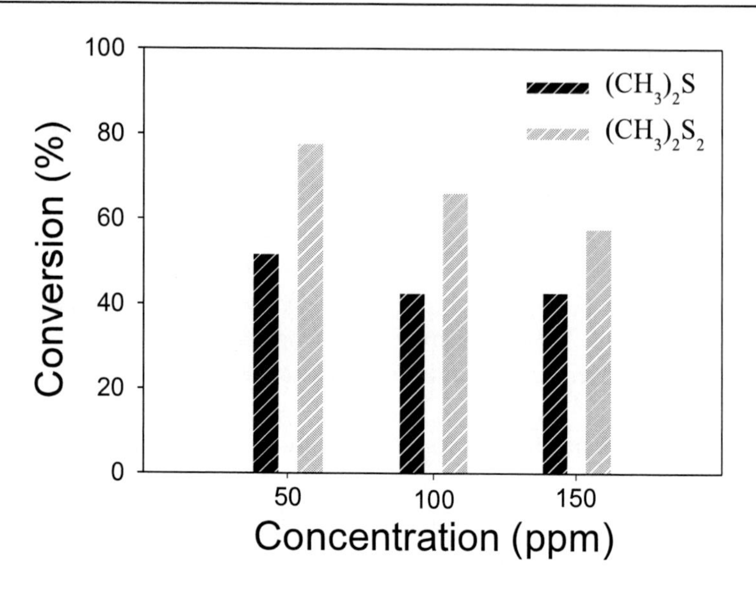

Figure 6. The effect of VOCs concentration on the catalytic conversion of $(CH_3)_2S$ and $(CH_3)_2S_2$ over a MnO/Fe_2O_3 catalyst. (temperature: 300°C; space velocity: 55,000 h^{-1}; O_2 concentration: 20.8%).

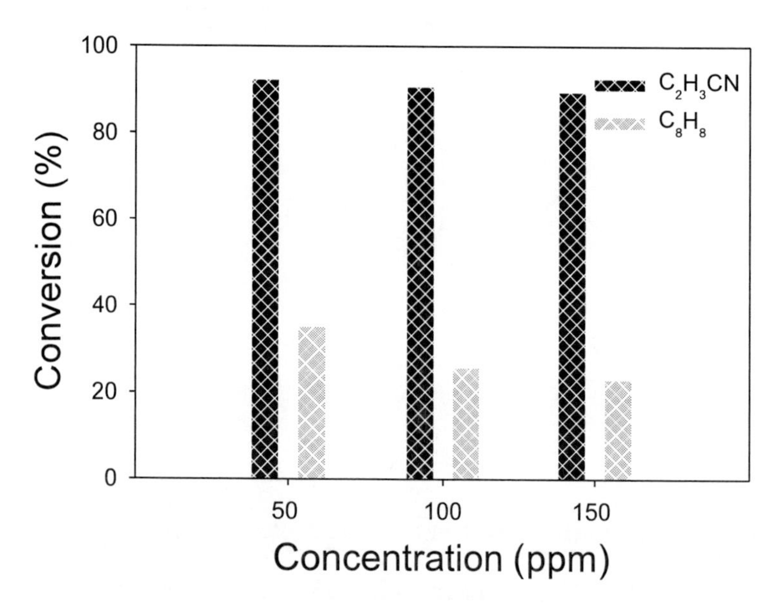

Figure 7. The effect of VOCs concentration on the catalytic conversion of C_2H_3CN and C_8C_8 over a MnO/Fe_2O_3 catalyst. (inlet temperature: 200°C; space velocity: 70,000 h^{-1}; O_2 concentration: 20.8%).

3.4. VOCs Concentration

The effects of $(CH_3)_2S$ and $(CH_3)_2S_2$ inlet concentrations on their conversions by the MnO/Fe_2O_3 catalyst are shown in Figure 6 (Chu et al., 2001a). It shows that the conversion of $(CH_3)_2S$ decreases as its concentration increases from 50 to 150 ppm. This result is consistent with the results of Chu et al., who used a Pt/Al_2O_3 catalyst to convert $(CH_3)_2S$ and CH_3SH (Chu et al., 2001b). Electron spectroscopy for chemical analysis (ESCA) of the fresh catalyst and the poisoned catalyst, used for incinerating 150 ppm CH_3SH at 315°C, 20.8% O_2 and 50,000 h^{-1} for 32 h, was conducted in their study. Their results showed that the sulfur content on the surface of the fresh catalyst and the poisoned catalyst were 0.2 and 2.8%, respectively. These data suggest that the activated site of the catalyst in this study may also be covered by sulfur compounds especially for the high sulfur content chemical. Figure 6 also shows the effect of $(CH3)_2S_2$ concentration on its conversion by the MnO/Fe_2O_3 catalyst. The concentration effects are similar to those for $(CH_3)_2S$.

Figure 7 shows the effect of C_2H_3CN and C_8C_8 concentrations on their conversions by the Mn_2O_3/Fe_2O_3 catalyst. It shows that the conversion of C_8H_8 decreases as its concentration increases from 50 to 150 ppm and suggests that organic compounds may cover the activated site of the catalyst, especially for the high concentration cases. The same phenomenon is also observed for C_2H_3CN. It can be found that the conversion of C_2H_3CN decreases as its concentration increases from 50 to 150 ppm. It suggests that the activated site of the catalyst may be covered by nitrogen compounds especially for high nitrogen content conditions. This is consistent with the result of Chu and Wu (Chu and Wu, 1998). Therefore, it suggests that the activated site of the catalyst may be covered by pollutant compounds especially for the high concentration conditions.

3.5. Space Velocity

Figure 8 shows that the lower the space velocity, the higher the conversions of $(CH_3)_2S$ and $(CH_3)_2S_2$ are. Figure 9 also shows that the lower the space velocity is, the higher the conversions of C_2H_3CN and C_8C_8 are. However, the differences are not significant for C_2H_3CN. It may be because the reaction rate of C_2H_3CN is very fast and the common space velocities used in the industry, which were used in this study, are all slow enough to finish the reaction.

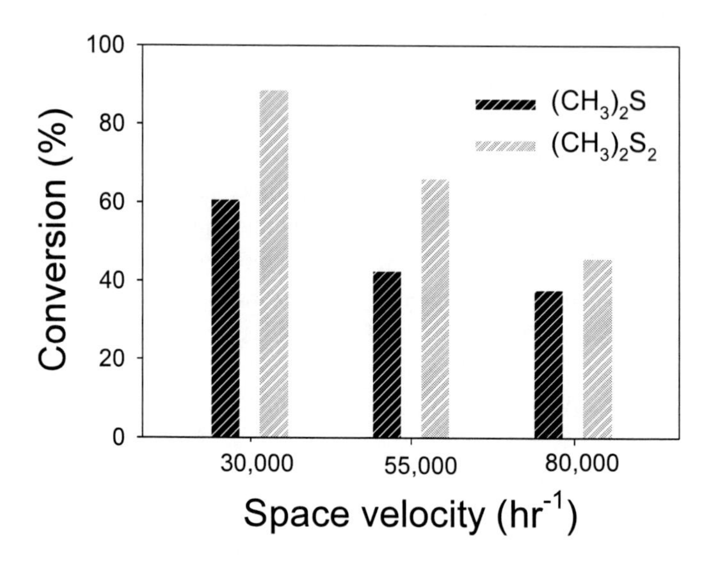

Figure 8. The effect of space velocity on the catalytic conversion of $(CH_3)_2S$ and $(CH_3)_2S_2$ over a MnO/Fe_2O_3 catalyst. (inlet temperature: 300°C; O_2 concentration: 20.8%).

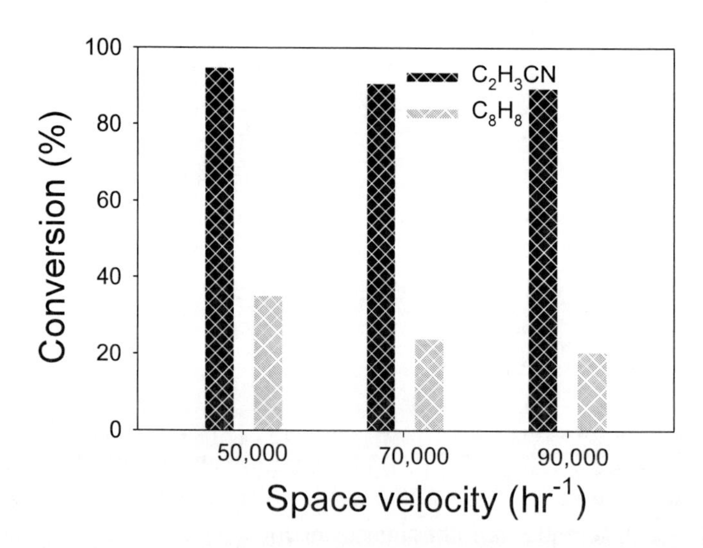

Figure 9. The effect of space velocity on the catalytic conversion of C_2H_3CN and C_8C_8 over a MnO/Fe_2O_3 catalyst. (inlet temperature: 200°C; O_2 concentration: 20.8%).

3.6. O_2 Concentration

The effects of O_2 concentration on the conversions of $(CH_3)_2S$ and $(CH_3)_2S_2$ are shown in Figure 10. It is found that the effect of O_2 concentration is not significant in the range of 0.1–20.8%. This result may be related to the oxygen atoms supplied by the MnO/Fe_2O_3 catalyst itself. On the other hand, the effects of O_2 concentration on the conversions of C_2H_3CN and C_8C_8 are shown in Figure 11. The results show that the O_2 concentration has a positive effect on the conversions of C_2H_3CN and C_8C_8. Ross and Sood (Ross and Sood, 1977) used $CoMO_4 \cdot H_2O$ to control a simulated pulp mill effluent gas containing methyl mercaptan, while O_2 concentrations varied from 1 to 4%. They found that the activity of the catalyst with respect to the production of an intermediate- dimethyl disulfide fell as the O_2 concentration in the effluent increased from 1 to 4%, while the production of SO_2, the complete oxidation product of methyl mercaptan, simultaneously increased.

4. Poisoning

A catalyst poison is an impurity present in the feed stream or in the process equipment that deposits onto the catalyst surface to reduce catalyst activity. There are two basic mechanisms by which poisoning occurs: (1) selective poisoning, in which a chemical directly reacts with the active site or the carrier and renders it less active or completely inactive, and (2) nonselective poisoning such as deposition of fouling agents onto or into the catalyst carrier, masking sites, and pores, which results in a performance loss caused by decreased accessibility of reactants to active sites.

Figure 12 shows a discriminating process by which a poison reacts directly with an active site, decreasing its activity or selectivity for a given reaction. Some poisons merely adsorb onto sites and block that site from further reaction.

These mechanisms are reversible in that heat treatment, washing, or simply removing the poison from the process stream often desorbs the poison from the catalytic site and restores its catalytic activity. When active sites are directly poisoned, there is a shift to high temperatures but with no change in the slope. This is because the remaining sites can function as before with no change in activation energy.

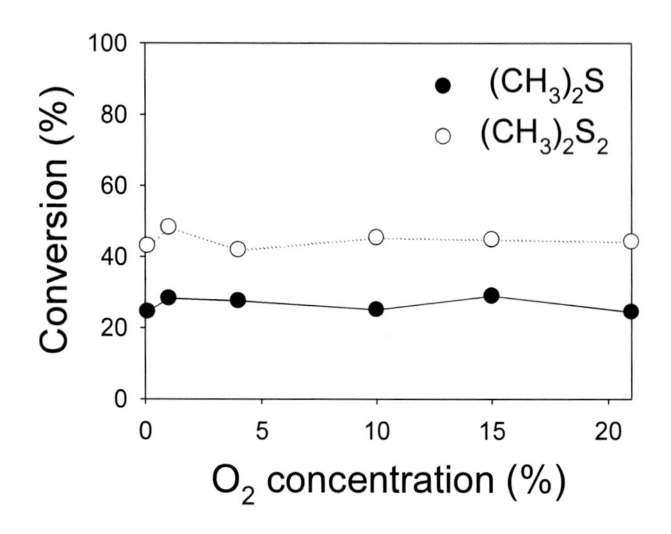

Figure 10. The effect of O_2 concentration on the catalytic conversion of $(CH_3)_2S$ and $(CH_3)_2S_2$ over a MnO/Fe_2O_3 catalyst (inlet temperature: 250°C; space velocity: 55,000 h^{-1}; O_2 concentration: 20.8%; VOCs concentration: 100 ppm).

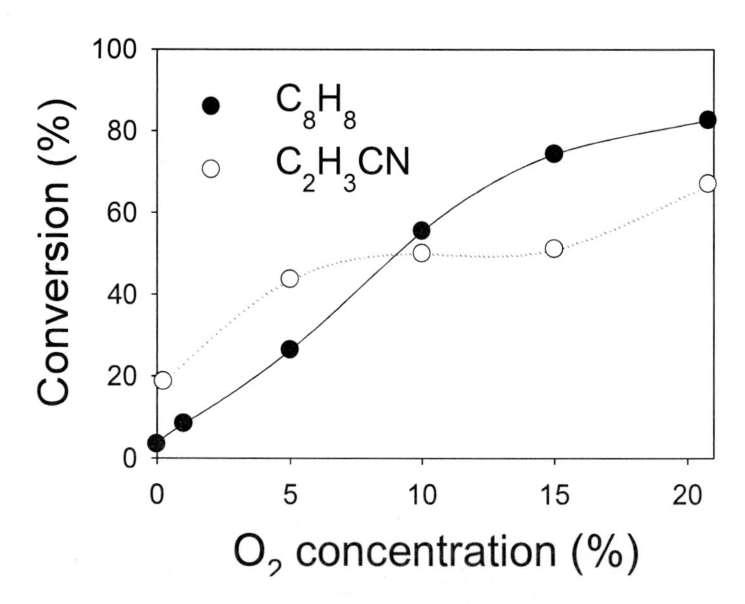

Figure 11. The effect of O_2 concentration on the catalytic conversion of C_2H_3CN and C_8C_8 over a MnO/Fe_2O_3 catalyst (C_2H_3CN inlet temperature: 150°C; C_8C_8 inlet temperature: 215°C; space velocity: 70,000 h^{-1}; O_2 concentration: 20.8%; VOCs concentration: 100 ppm).

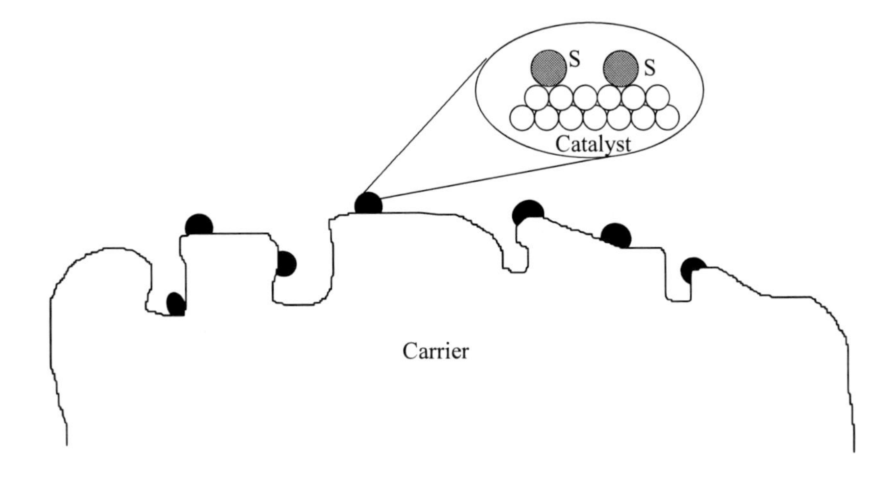

Figure 12. Conceptual diagram showing of the poisoned sites.

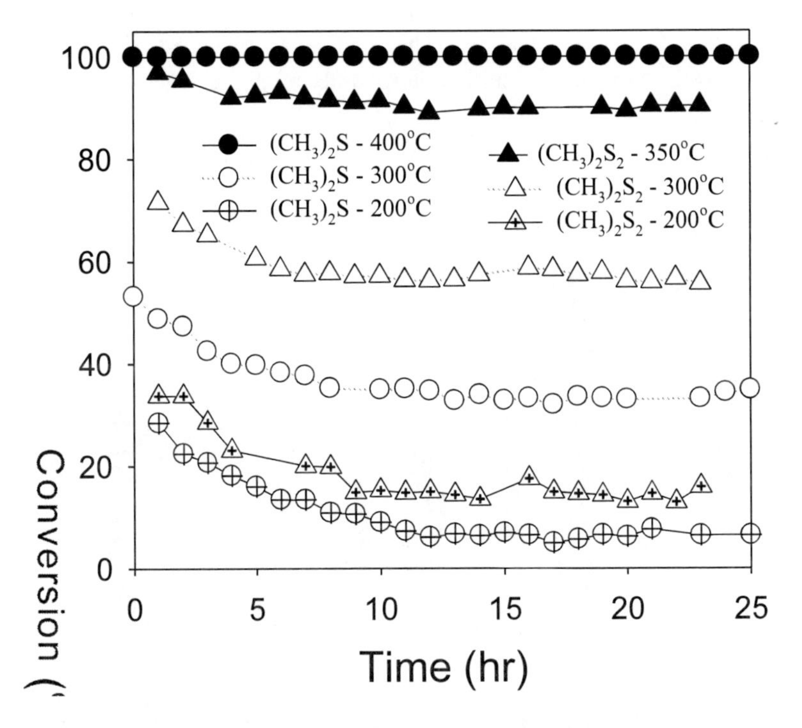

Figure 13. Poisoning effect of $(CH_3)_2S$ and $(CH_3)_2S_2$ on the MnO/Fe_2O_3 catalyst: $(CH_3)_2S$ and $(CH_3)_2S_2$. ($(CH_3)_2S$ inlet temperatures: 400, 300 and 200°C; $(CH_3)_2S_2$ inlet temperatures: 350, 300 and 200°C; VOCs concentration: 100 ppm; space velocity: 55,000 h^{-1}; O_2 concentration: 20.8%).

Life-tests of the MnO/Fe_2O_3 catalyst were conducted utilizing 100 ppm of $(CH_3)_2S$ and$(CH_3)_2S_2$ to identify the poisoning effect of sulfur-containing VOCs on the MnO/Fe_2O_3 catalyst. The results are shown in Figure 13. For temperatures below 300°C, the performance of the catalyst declines dramatically and then stabilizes. This phenomenon may be due to certain activated sites on the catalyst forming irreversible sulfur-poisoned sites; this process would need time to accomplish the irreversible reaction. The remaining sites of the catalyst could be reversible sulfur-poisoned sites. At 400◦C, however, no sulfur poisoning effect is found. This result is consistent with the results of Chu et al. (Chu and Wu, 1998; Chu et al., 2001a, 2001b) , who showed that $(CH_3)_2S$, $(CH_3)_2S_2$ and C_2H_5SH had a poisoning effect on the Pt/Al_2O_3 catalyst and the MnO/Fe_2O_3 catalyst at lower temperatures. Figure 14 shows the specific surface areas of the MnO/Fe_2O_3 catalyst for 25-h dimethyl sulfide poisoning at various temperatures (200, 300, and 400°C) and for 24-h dimethyl disulfide poisoning at various temperatures (200, 300, and 350°C). The results show that the specific surface areas of the poisoned catalysts are less than that of the fresh catalyst. This finding also suggests that the catalyst surface may be covered by sulfur compounds after the reaction. The surface area differences between the fresh catalyst and the poisoned catalyst for the case of $(CH_3)_2S$ are more dramatic than that of $(CH_3)_2S_2$.

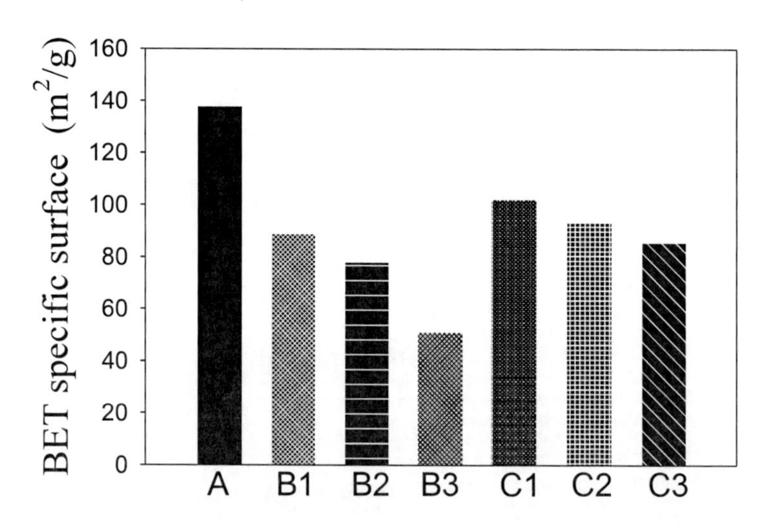

Figure 14. Specific surface area of the fresh catalyst with $(CH_3)_2S$ poisoning for 25 h and $(CH_3)_2S_2$ poisoning for 24 h at various temperatures. (A: fresh catalyst; B1: $(CH_3)_2S$-200°C; B2: $(CH_3)_2S$-300°C; B3: $(CH_3)_2S$-400°C; C1: $(CH_3)_2S_2$-200°C; C2: $(CH_3)_2S_2$-300°C; and C3: $(CH_3)_2S_2$-350°C).

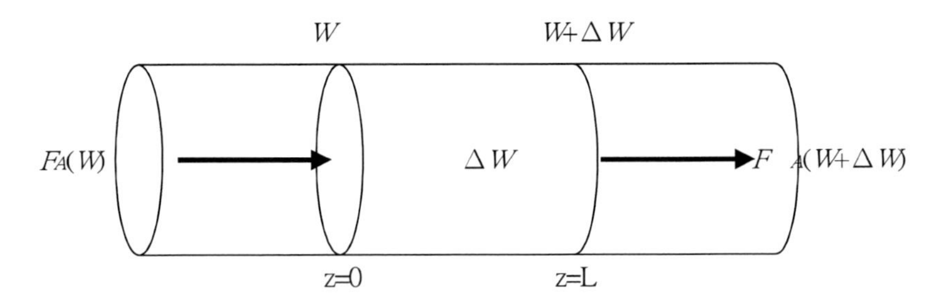

Figure 15. The scheme of a plug–flow reactor.

5. KINETIC STUDY

In the present work, the Packed–Bed Plug–Flow reactor design was used for best fit of kinetic models. In differential operation, the conversion from the reactor is kept below 5~10%.

5.1. Differential Reactor Operation

The Plug–Flow reactor (PFR) might initially have a square, circular, triangular or other cross–section shape. To develop the mole balance equation, we consider a small weight element, ΔW, of the reactor, as shown in Figure 15.

As with the PFR, the Packed–Bed reactor (PBR) is assumed to have no radial gradients in concentration, temperature, or reaction rate. The generalized mole balance on species A over catalyst weight, ΔW, was shown below.

$$in - out + generation = accumulation$$

$$F_A(W) - F_A(W + \Delta W) + (-r)\Delta W = 0$$

After dividing by ΔW and taking the limit as $\Delta W \rightarrow 0$, we arrive at the differential form of the mole balance for a Packed–Bed reactor.

$$-r = \frac{dF_A}{dW} \tag{1}$$

For differential reactor operation it is assumed that the concentration change is small enough so that the plug flow equation can be approximated by:

$$-\frac{(F_{A,L}-F_{A,0})}{W}=\frac{F_{A,0}X_A}{W}=(-r) \qquad (2)$$

where X_A is the fractional conversion of A, and $F_{A,0}$ and $F_{A,L}$ represent the molar flow rates of A at z= 0 and z= L, respectively (z is the axial distance and $F_{A,L}= F_{A,0}$ (1 – X_A). Provided that the inlet and outlet flow rates and concentrations are known, the rate can be easily calculated. The rate is assumed to correspond to the average of the inlet and outlet concentrations. Ease of data analysis is an advantage of the differential method of operation. Another advantage is that, because the conversion is low, the amount of heat liberated is also relatively small. This helps in maintaining isothermal operation, which is especially difficult to achieve with highly exothermic combustion reactions. Two drawbacks to the method are:

A. The extreme accuracy and precision required in the parameter measurement to keep the errors within acceptable limits,
B. Few products are produced, so separate experiments must be performed to evaluate the effects of products on the kinetics. It can be quite time consuming to prepare the large number of "synthetic feeds" (feeds with reactants and products present) required to obtain sufficient data to elucidate the effects of products.

Each experiment (that is, *any* one given set of experimental conditions) gives a single value for the rate of reaction at the given temperature and concentration. The appropriate concentration to use is the average of the inlet and the outlet values. In order to gather sufficient data to determine the parameters in the rate expression, a large number of experiments may have to be performed.

5.2. Power–rate Law

For studies of homogeneous gas–phase reactions, the rate of an elementary bimolecular reaction between two species, A and B, is given by (Satterfield, 1980)

$$\text{Rate} = \frac{molecules reacted}{(time)(volume)} = kC_A C_B = A_l^{-E/RT} C_A C_B \quad (3)$$

Equation (3) is known as the Arrhenius expression when A, the pre-exponential factor, is taken to be independent of temperature. Since in any event the effect of temperature on A is small relative to its effect on the exponential term, one may with little error take A to be independent of temperature.

By analogy a simple expression for the rate –r of a heterogeneous catalyzed reaction between A and B is

$$\text{Rate} = \frac{molecules reacted}{(time)(volume)} = k_0 e^{-E/RT} f(C_A C_B) \quad (4)$$

Where k_0 is taken to be independent of temperature and surface area of the catalyst. The function of the concentrations, which usually is easiest to use in correlating rate data, consists of simple power functions: C_A^a, C_B^b, where a and b are empirically adjusted constants, hence

$$-r = k_0 e^{-E/RT} C_A^a C_B^b \quad (5)$$

The term k_0 will usually have no theoretical significance, and the exponents may be integral or fractional, positive, zero, or negative.

Many catalytic reactions follow a simple relationship of this type over a sufficiently wide range of conditions as to make the correlation useful. Although the power–rate law is a simple and quite useful tool for simulating most reactors, a more confusing series of reacting systems are also frequently found in which the concentration function in equation (5) remains constant but e and k_0 both change.

The apparent order of the reaction and the apparent activation energy may change with temperature, which requires developing a different kinetic model, such as the Mars and van Krevelen model and Langmuir–Hinshelwood model discussed bellow.

5.3. Mars and van Krevelen Model

The behavior of most oxidation catalysts can be interpreted within the framework of a redox mechanism. The hypothesis of this mechanism consists of two steps:

A. Reaction between catalyst in an oxidized form, *Cat–O*, and the hydro carbon *R*, in which the oxide becomes reduced:

$$Cat - O + R \xrightarrow{k_i} RO + Cat - \tag{6}$$

B. The reduced catalysts, *Cat –*, becomes oxidized again by oxygen from the gas phase:

$$2Cat - + O_2 \xrightarrow{k_o} 2Cat - O \tag{7}$$

The reaction rates of the two steps should be equivalent at the steady state.

To develop a simple mathematic model, Mars and van Krevelen assumed the oxidation rate of VOCi, –ri, to be proportional to the percentage of activated sites of the oxidized catalyst, θ and VOC_i concentration, C_i and the oxidation rate of the reduced catalyst, $-r_O$, to be proportional to the percentage of reduced sites of the catalyst, $1-\theta$ and O_2 concentration, C_O. This can be shown as:

$$-r_i = k_i C_i^m \theta \tag{8}$$

$$-r_O = k_O C_O^n (1-\theta) \tag{9}$$

If α models of O_2 are required to oxidize 1 mol of VOC_i, $-r_O$ should be equivalent to $-\alpha r_i$ at the steady state. The percentage of activated sites of the oxidized catalyst, θ, can be rearranged to (Gangwal, et al., 1988):

$$\theta = \frac{k_O C_O^n}{k_O C_O^n + \alpha k_i C_i^m} \tag{10}$$

Substituting equation (10) into the equation (8) gives

$$\frac{1}{-r} = \frac{\alpha}{k_O C_O^n} + \frac{1}{k_i C_i^m} \quad (11)$$

Equation (11) can be simplified to two different functions.

(a) Fix–oxygen state, the C_O term can be assumed as a constant. Therefore, equation (11) can be illustrated as a linear relation between $1/(-r)$ and $1/C_i^m$ at various m.
(b) Fix–VOC state, the C_i term can be assumed as a constant. Equation (11) can be illustrated as a linear relation between $1/(-r)$ and $1/C_O^n$ at various n.

5.4. Langmuir–Hinshelwood Model

The assumptions underlying the Langmuir adsorption isotherm are retained. Further, adsorption equilibrium is assumed to be established at all times; for example, the rate of reaction is taken to be much less than the potential rate of adsorption or desorption. The concentrations of adsorbed species are therefore determined by adsorption equilibrium as given by the Langmuir isotherm. If two or more species are present, they compete with each other for adsorption on a fixed number of active sites.

Reaction is assumed to occur between adsorbed species on the catalyst. If two reactants are decomposed, the process may be assumed to be either molecular or bimolecular, depending upon the number of product molecules formed per reactant molecule and whether or not the products are adsorbed. In this study, we assume that the products are not adsorbed on the catalysts surface. The rate expression can be derived for one postulated mechanism of the adsorption of two gases on the same type active site. The form and complexity of the expression depend on the assumptions made concerning this mechanism and presented below:

$VOC_i + O_2 \rightarrow$ products

Assume:

A. VOC_i and O_2 both may be appreciably adsorbed.

B. The reaction rate is proportional to the quantity of adsorbed VOC_i and O_2 molecules. Then

$$-r,\frac{(moles)}{(time)(area)}=k\theta_i\theta_O \tag{12}$$

C. No dissociation of VOC_i and O_2 molecules occurs on adsorption.

D. Reverse reaction is negligible.

Using the Langmuir adsorption isotherm, the fraction of surface covered by VOC_i and O_2 can be derived as follows:

$$\theta_i = K_iC_i[1-(\theta_i+\theta_O)] = K_iC_i(1-\sum\theta) \tag{13}$$

$$\theta_O = K_OC_O[1-(\theta_i+\theta_O)] = K_OC_O(1-\sum\theta) \tag{14}$$

Adding equations (13) and (14),

$$(1-\sum\theta)=\frac{1}{1+K_iC_i+K_OC_O} \tag{15}$$

Since two molecules are formed for each one that reacts, and it is postulated that both product molecules are adsorbed, it would seem plausible that it is necessary for an empty site to be present adjacent to the reacting molecule to accommodate one of the product molecules. Combine equations (12), (13), (14), and (15) gives

$$-r=\frac{kK_iC_iK_OC_O}{(1+K_iC_i+K_OC_O)^2} \tag{16}$$

Assume that O_2 dissociates upon adsorption and associates on desorption. In order for dissociation to occur, a gas molecule must plausibly impinge on the surface at a location where two sites are adjacent to one another. Up to fairly high fractional coverage, the number of pairs of adjacent sites is proportional to the square of the number of single sites. Then the fraction of surface covered by O_2 is given by

$$\theta_O = \sqrt{K_O C_O}\left[1-(\theta_i+\theta_O)\right] = \sqrt{K_O C_O}(1-\sum\theta) \tag{17}$$

Equation (12) can be rewritten as follow:

$$-r = \frac{kK_i C_i \sqrt{K_O C_O}}{(1+K_i C_i + \sqrt{K_O C_O})^2} \tag{18}$$

Several kinetic studies of the catalytic incineration of CH_3SH and $(CH_3)_2S$ over a Pt/Al_2O_3 catalyst (Chu and Horng, 1998), $(CH_3)_2S_2$ over a Pt/Al_2O_3 catalyst (Chu et al., 2001a), $(CH_3)_2S_2$ over an MnO/Fe_2O_3 catalyst (Chu et al., 2001c), and C_8H_8 over an MnO/Fe_2O_3 catalyst (Tseng and Chu, 2001) were carried out in a bench scale catalytic incinerator, respectively. A differential reactor design was used for the system. Three kinetic models, such as the power-rate law, the Mars and van Krevelen model and the Langmuir-Hinshelwood model were used to best fit the experimental data.

Tseng and Chu (2001) used the MnO/Fe_2O_3 catalyst to catalytically convert styrene in a differential reactor. A kinetic experiment was performed to get the rate constant, k, for the Langmuir-Hinshelwood model with molecular O_2 adsorption and atomic O adsorption. The results indicate that the activation energy for the case of molecular O_2 adsorption, Ea=28.0 kJ/mol, is in the proper range for the catalytic reaction. This is consistent with the results of the previous study (Chu and Wu, 1998), in which the results showed that the adsorption of O_2 molecule is important in the process of catalytic incineration of CH_3SH described by the Langmuir-Hinshelwood model. For the case of atomic O adsorption, the result shows that the activation energy for the styrene catalytic decomposition reaction, Ea=29.2 kJ/mol, is also in the proper range for the catalytic reaction. Therefore, both the Langmuir-Hinshelwood models involving molecular O_2 adsorption and atomic O adsorption are feasible to describe the catalytic incineration of C_8H_8 over the MnO/Fe_2O_3 catalyst. Chu and Horng (1998) also showed a similar phenomenon that the Langmuir-Hinshelwood model is suitable for the catalytic incineration of both CH_3SH and $(CH_3)_2S$.

Chu et al. (2001) used the MnO/Fe_2O_3 catalyst to catalytically convert $(CH_3)_2S$ and $(CH_3)_2S_2$ in a differential reactor. The result suggests that the Langmuir-Hinshelwood model may be a feasible method of describing the catalytic incineration of $(CH_3)_2S$ over the MnO/Fe_2O_3 catalyst. The results also indicate that the Langmuir-Hinshelwood model may also be a good method to describe the catalytic incineration of $(CH_3)_2S_2$ at low temperatures form 70°C to 130°C.

5.5. Verification of the Kinetic Model

To further establish the fitness of the three kinetic models, we have predicted the conversion of VOC_i on the isothermal integral packed–bed reactor at various temperatures by the three kinetic models.

In the prediction of the conversion of VOC_i on the isothermal integral packed–bed reactor at various temperatures, equation (1) can be rearranged to an ordinary differential equation:

$$\frac{dW}{dX} = \frac{F}{(-r)} \tag{19}$$

Furthermore, equation (19) can be solved by an integration method. By rearranging equation (19) and integrating both sides, we can obtain the follow equation:

$$W = \int dW = F \times \int \frac{dX}{(-r)} \tag{20}$$

Further coupling with the three kinetic models gives:

A. Mars and van Krevelen model:

$$W = F \times \int \frac{\alpha k_i C_{in}(1-X) + k_O C_O}{k_i k_O C_O C_{in}(1-X)} dX \tag{21}$$

B. Langmuir–Hinshelwood model:

(1) Molecular Oxygen Adsorption:

$$W = F \times \int \frac{(1 + K_O C_O + K_i C_{in}(1-X))^2}{k K_O C_O K_i C_{in}(1-X)} dX \tag{22}$$

(2) Atomic Oxygen Adsorption:

$$W = F \times \int \frac{(1+\sqrt{K_O C_O} + K_i C_{in}(1-X))^2}{k\sqrt{K_O C_O} K_i C_{in}(1-X)} dX \tag{23}$$

C. Power rate law:

$$W = F \times \int \frac{1}{k C_O^b C_{in}^a (1-X)^a} dX \tag{24}$$

The Newton–Cotes quadrate formula was employed to solve the conversion, *X*. In the integration, *X* was increased step by step until the value of *W* (weight of used catalyst) was attained.

Tseng et al. (2005) used the values obtained by the three models to compare with the experimental data of catalytic incineration of methyl isobutyl ketone over a Pt/γ-Al_2O_3 obtained in the isothermal integral fixed-bed reactor (W=0.6 g). The result indicated that the power-rate law was not feasible for describing the catalytic incineration of methyl isobutyl ketone over the Pt/γ-Al_2O_3 catalyst.

The predicted values by Eq. (23) did not fit the conversion of methyl isobutyl ketone. This suggested that both the power-rate law and the Langmuir–Hinshelwood model (atomic oxygen adsorption) were not feasible for describing the catalytic incineration of methyl isobutyl ketone over the Pt/γ-Al_2O_3 catalyst.

They also compared the predicted values by equations (21) and (22) and experimental data obtained by an isothermal integral fixed-bed reactor at temperatures ranging from 373 to 573 K. It revealed that the experiment fitted the predicted values by equations (21) and (22). The results indicated that the kinetic mechanism would be the molecular oxygen adsorption effect and the redox mechanism on the surface of the catalyst.

This suggested that both the Mars and van Krevelen model and Langmuir–Hinshelwood model (molecular oxygen adsorption) were feasible for describing the catalytic incineration of MIBK over the Pt/γ-Al_2O_3 catalyst.

REFERENCES

Agarwal, S. K., Spivey, J. J., Butt, J. B. Catalyst deactivation during deep oxidation of chlorohydrocarbons. *Appl. Catal. A: General* 1992, 82, 259–275.

Artizzu–Duart, P., Brullé, Y., Gaillard, F., Garbowski, E., Guilhaume, N., Primet, M. Catalytic combustion of methane over copper– and manganese–substituted barium nexaaluminates. *Catal. Today* 1999, 54, 181–190.

Bickle, G. M., Suzuki, T., Mitarai, Y. Catalytic destruction of chlorofluorocarbons and toxic chlorinated hydrocarbons. *Appl. Catal.* B 1994, 4, 141–153.

Chandler, B.D., Schabel, A.B., Pignolet, L.H. Preparation and Characterization of Supported Bimetallic Pt-Au and Pt-Cu Catalysts from Bimetallic Molecular Precursors. *J. Catal.* 2000, 193, 186-198.

Chu, H., Horng, K., The kinetic of catalytic incineration of CH_3SH and $(CH_3)_2S$ over a Pt/Al_2O_3 catalyst. *Sci. Total Environ.* 1998, 209, 149–156.

Chu, H., Lee, W.T., Chiou, Y.Y., Tseng, T.K. The Kinetics of Catalytic Incineration of C_2H_5SH and $(CH_3)_2S_2$ over a Pt/Al_2O_3 Catalyst; *Environ. Technol.* 2001a, 22, 515-522.

Chu, H., Lee, W.T., Horng, K.H., Tseng, T.K. The catalytic incineration of $(CH_3)_2S$ and its mixture with CH_3SH over a Pt/Al_2O_3 catalyst. *J. Hazard. Mater.* 2001b, 82, 43-53.

Chu, H., Wu, L.W. The catalytic incineration of ethyl mercaptan over a MnO/Fe2O3 catalyst. *J. Environ. Sci. Health.* Part A Toxic/Hazard. *Subst. Environ. Eng.* 1998, 33, 1119-1148.

Chu, H., Hao, G.H., Tseng, T.K., The Kinetics of Catalytic Incineration of Dimethyl Sulfide and Dimethyl Disulfide over an MnO/Fe_2O_3 Catalyst. *J. Air & Waste Manage. Assoc.* 2001c, 51, 574-581.

de Nevers, N. Air Pollution Control Engineering, 2nd ed. McGraw–Hill, *Singapore,* 2000; 571–573.

Drago, R. S., Jurczyk, K., Singh, D. L., Young, V. Low–temperature deep oxidation of hydrocarbons by metal oxides supported on carbonaceous materials. *Appl. Catal. B: Environ.* 1995, 6, 155–168.

Gangwal, S.K., Mullins, M.E., Spivey, J.J., Caffrey, P.R. Kinetics and selectivity of catalytic oxidation of n–hexane and benzene. *Appl. Catal.* 1988, 36, 231–247.

Heyes, C.J., Irwin, J.G., Johnson, H.A., Moss, R.L. The catalytic oxidation of organic air pollutants, part 1. single metal oxide catalysts. *J Chem. Technol. Biotechnol.* 1982, 32, 1025-1033.

Kang, Y. M.; Wan, B. Z. Effects of acid or base additives on the catalytic combustion activity of chromium and cobalt oxides. *Appl. Catal. A: General* 1994, 114, 35–49.

Kominami, H., Takada, Y., Yamagiwa, H., Kera, Y., Inoue, M., Inui, T. Synthesis of thermally stable nanocrystalline anatase by high-temperature hydrolysis of titanium alkoxide with water dissolved in organic solvent from gas phase. *J. Mater. Sci. Lett.* 1996, 15, 197-200.

Komiyama, M. Design and preparation of impregnated catalysis. Catalysis Reviews:. *Sci. Eng.* 1985, 27, 342-372.

Ross, R.A., Sood, S.P. Catalytic oxidation of methyl mercaptan over cobalt molybdate. *Ind. & Eng. Chem. Prod. Res. Dev.* 1977, 16, 147-150.

Rossin, J. A., Farris, M. M. Catalytic oxidation of chloroform over a 2% platinum alumina catalyst. *Ind. Eng. Chem. Res.* 1993, 32, 1024–1029.

Satterfield, C.N. *Heterogeneous catalysis in practice.* McGraw–Hill: New York, 1980, 86–92.

Spivey, J.J. Complete catalytic oxidation of volatile organics. *Ind. & Eng. Chem. Res.* 1987, 26, 2165-2180.

Stiles, A. Catalyst manufacture. *Laboratory and commercial preparations.* Marcel Dekker: New York, 1983.

Taylor, S.H., Heneghan, C.S., Hutchings, G.J., Hudson, I.D. The activity and mechanism of uranium oxide catalysts for the oxidative destruction of volatile organic compounds. *Catal. Today* 2000, 59, 249–259.

Thomas, A.H., Brundrett, C.P. Catalyst development: Lab to commercial scale. *Chem. Eng. Prog.* 1980, 76, 41-45.

Tichenor, B.A., Palazzolo, M.A. Destruction of volatile organic compound via catalytic incineration. *Environ. Prog.* 1987, 6, 172-176.

Toledo, J.M., Corella, J., Sanz, A., Noble metal–based catalysts for total oxidation of chlorinated hydrocarbons. *Environ. Prog.* 2001, 20, 167–174,.

Tseng, T.K., Chu, H. The kinetic of catalytic incineration of styrene over a MnO/Fe_2O_3 catalyst. *Sci. Total Environ.* 2001, 275, 83–93.

Tseng, T.K., Chu, H., Ko, T.H., Chaung, L.K., The kinetic of the catalytic decomposition of methyl isobutyl ketone over a $Pt/\gamma\text{-}Al_2O_3$ catalyst. *Chemosphere.* 2005, 61, 469–477.

van der Vaart, D. R., Vatavuk, W. M., Wehe, A. H. Thermal and catalytic incinerators for the control of VOCs. *Journal of Air & Waste Management Association* 1991a, 41(1), 92–98.

van der Vaart, D. R., Vatavuk, W. M., Wehe, A. H. The cost estimation of thermal and catalytic incinerators for the control of VOCs. *J. Air & Waste Manage. Assoc.* 1991b, 41(4), 497–501.

Völter, J., Lietz, G., Spindler, H., Lieske, H. Role of metallic and oxidic platinum in the catalytic combustion of n-heptane. *J. Catal. 1987, 104, 375-380.*

Ward, J.W. The nature of active sites on zeolites: 1. The decationated Y zeolites. *J. Catal.* 1967, 9, 225.

Watanabe, N., Yamashita, H., Miyadera, H., Tominaga, S. Removal of unpleasant odor gases using an Ag–Mn catalyst. *Appl. Catal. B: Environ.* 1996, *8*, 405–415.

Worstell, J.H. Succeed at catalyst upgrading. *Chem. Eng. Prog.* 1992, 88, 33-39.

In: Volatile Organic Compounds ISBN 978-1-61324-156-1
Editors: J. C. Hanks et al. pp. 119-147

Chapter 4

TRANSPORT OF VOCS IN POLYMERS

***Karel Friess*[1], *Pavel Izák*[2], *Milan Šípek*[1] and *Johannes Carolus Jansen*[3]**

[1]Institute of Chemical Technology,
Technická 5, 160 00 Prague 6 - Dejvice, Czech Republic
[2]Institute of Chemical Process Fundamentals,
Rozvojová 135, 165 02 Prague 6 - Sedlec, Czech Republic
[3]Institute on Membrane Technology, ITM-CNR,
Via P. Bucci 17/C, 87030 Rende (CS), Italy

1. ABSTRACT

Leaking of volatile organic compounds (VOCs) from gasoline during its storage, handling and transportation constitutes a serious ecological problem seeing that VOCs are known as toxic, environmentally harmful and carcinogenic agents. Despite the fact that the lost amounts of mainly hydrocarbons during common operations in refineries or at fuel stations may seem negligible; in reality they reach hundreds of tons per year of valuable industrial products. All above mentioned facts are the reason why the separation of these compounds from air and their recycling is critically important. At the present time, VOCs removal from the air is realized by traditional cost-consuming technologies like adsorption or refrigeration and by ecological progressive high-efficiency membrane separations. Polymer membranes based on polydimethylsiloxane (PDMS) or polyether-block-amide (PEBA) currently belong to the group of polymers used for the preparation of composite separation membranes [1-

4]. Some of their unfavourable limitations (lower chemical resistance, swelling) led researchers to test also other potentially utilisable polymers like polyvinylidene fluoride (PVDF), poly{1-trimythylsilyl-1-propyne} (PTMSP), high free volume amorphous glassy perfluoropolymers (Teflon AF) or cross-linked poly(amide-imide) polymers[2-4]. Hence, detailed knowledge of polymer structure-permeability relationship and polymer-penetrant interactions plays an important role in the development and potential industrial application of newly prepared membrane materials. Generally, the mass transport of VOCs in and through the polymer matrix is a complex process which depends on polymer properties (glassy/ rubbery state, orientation, porous/nonporous structure, symmetric/ asymmetric architecture etc.), penetrant properties (molecular size and shape, specific penetrant-penetrant or penetrant-polymer interactions) and also on external conditions (temperature, pressure, concentration gradient etc.). It is generally accepted that mass transport in dense polymer membranes takes place according to the well known solution-diffusion mechanism (SDM) [5]. For small, not-self-aggregative, low-sorbed molecules SDM is valid without any limitations. In other cases, especially for VOCs the solubility of the compound has a strong influence on the polymer behaviour (swelling, plasticization, chain flexibility, reorganisation of dynamic free volume elements) and, consequently, on diffusivity and permeability [1, 5]. Therefore the concentration-dependence of transport parameters must be taken into account. In this chapter we shall give a survey of the VOCs transport in non-porous polymer membranes with special reference to the phenomenon of concentration-dependence of transport.

2. Transport through Polymer Membranes

The transport mode of penetrant molecules in polymeric membranes depends on the relative size of the penetrant and the size and arrangement of the void structure in the membrane. It is fundamentally different for porous (from macroscopic to nanoscale) and for nonporous membranes.

Molecular-level based theories [6, 7] postulate the diffusion process in terms of polymer chain motion. Based on X-ray diffraction results the theory assumes that non-crystalline polymer regions possess an approximate semicrystalline order with chain bundles that are locally parallel along distances of several nanometers. In fact, the existence of order both along the polymer backbone and perpendicular to it has spurred the development of numerous “bundle models” which assume that regions of local order, characterized by various degrees of orientation order, exist in bulk amorphous materials [8]. The theory also assumes a specific chain packing and potentially distinct molecular motion in a “specific tube or channel” along the parallel polymer chains [6] or perpendicular to them [7]. The first process is relatively fast; it requires little activation energy and determines the effective jump length in diffusion. In contrast, the second process requires activation energy equal to the energy necessary to produce a minimum chain separation that will accommodate the molecule and permit its transfer. If both processes occur simultaneously, the first would predominate [8].

In case of macroscopic theories, the permeation of small gases or vapours molecules through porous membranes can be described by Knudsen diffusion and Poiseuille flow [9]. The proportions of Knudsen to Poiseuille flow are governed by the ratio of the pore radius (r) to the mean free path (λ) between the molecules in the gas phase. If ratio r/λ is < 1, then Knudsen flow-type predominates and molecule-to-pore wall collisions are more likely than molecule-to-molecule collisions. If the ratio r/λ increases, collisions among the molecules are more frequent and for $r/\lambda > 5$ Poiseuille flow predominates. Because the range of the mean free path of gases or vapours at atmospheric pressure is 100-200 nm, the membrane pore radius must be less than 50 nm to have predominantly Knudsen flow. With further decreasing pore size also the extent of Knudsen flow diminishes. The pore model of membrane transport is generally considered to be valid down to about 5 Ångström or less [9].

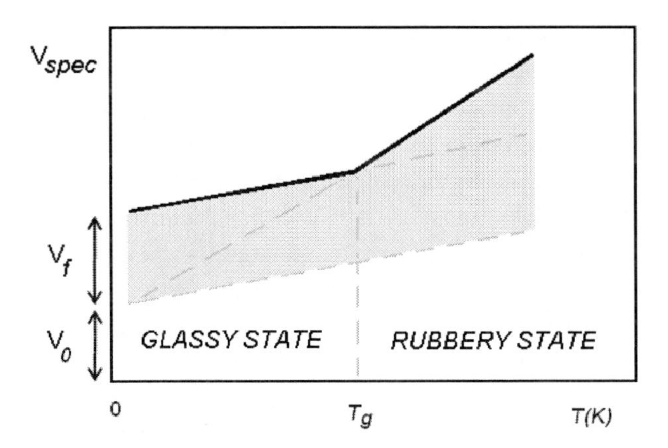

Figure 1. Free volume in polymer. The total (specific) volume of polymer membrane V_{spec} at given temperature T is sum of volume V_0 (volume occupied by molecules at T = 0 K) and free volume V_f (empty volume between the polymer chains). In polymer chemistry very often the fractional free volume V_{ffv} is used as the ratio of free volume and specific volume of a polymer ($V_{ffv} = V_f / V_{spec}$).

The mechanism of gas and vapour transport through nonporous membranes is fundamentally different from that in micro-porous membranes [1, 5]. It is generally accepted that the diffusion process in a nonporous polymer depends on the free volume (empty volume between the polymer chains) in the polymer matrix, regardless whether the polymer is above or below the glass transition temperature T_g (Figure 1).

Above the glass temperature the polymer behaves as a rubber, and the available free volume for penetrant transport increases with temperature; the penetrant solubility depends linearly on pressure and transport can be described by the solution-diffusion mechanism [5] containing permeability P, diffusion D and solubility(sorption) S coefficients

$$P = DS \tag{1}$$

Also below T_g the free volume increases with temperature. Below the glass transition temperature the polymer becomes a glass and its free volume is nearly independent of temperature and if aging effects are neglected it is fixed at a certain value which depends on its thermo-mechanical history, i.e. the rate at which it is cooled through the glass transition temperature in the case of melt-processing, or the evaporation and drying conditions in the case of solution casting. At relatively low pressure the situation is the same as in rubbers. Only at high pressure S and D become more complex because of the dual mode sorption with two concurrent sorption modes [10].

Based on SDM model, the mass transport process of gas and vapours can be described as a sequence of consecutive processes of sorption of the components on the feed side of the membrane surface, activated diffusion through membrane and desorption of the components on the permeate side of the membrane. The mass transport is caused by the driving force, which in most cases is the gradient of pressure, temperature or chemical potential. This mechanism has been in 1886 proposed by Thomas Graham[1], who studied the gas permeability of rubber materials. Later on Wroblewski[2] quantitatively described permeability in relation to Fick´s law and Henry´s law and his ideas are basically valid up to this day. Later studies of the structure of macromolecular substances essentially contributed to the theory of mass transport through the polymers and prediction of the permeability of polymers

[1] T. Graham, On the Molecular Mobility of Gases, Journal of the Chemical Society of London, 17 (1864) 334-339.

[2] S. Wroblewski, Ueber die Diffusion der Gase durch absorbirende Substanzen, Annalen der Physik, 234 (1876) 539–568.

in dependence on their composition, structure, phase state, orientation and other properties [1, 3, 4, 8, 9]

The free volume approach provides additional understanding of the segmental motions of polymers and, consequently, of the diffusion process inside a polymeric matrix [11, 12]. In the case of glassy polymers, the diffusion coefficients for penetrant-polymer mixtures were found to be history dependent for particular systems [13]. A larger amount of excess free volume is frozen into the polymer in the non-equilibrium glassy state if the polymer is rapidly quenched rather than slowly cooled. Consequently, it follows that if isothermal aging effects are negligible the penetrant diffusion coefficient D in polymer exposed to rapid cooling is higher than D in a slowly cooled polymer.

For vapours in glassy polymers, where distinct forms of interaction between penetrant and polymer take place, anomalous transport with a combination of Fickian case I and case II diffusion was reported for various polymer-penetrant systems [14, 15] and several correlations describing and predicting such form of transport were developed [16, 17]. It was recently suggested that the unusually broad transient in methanol vapour permeation could be related to the formation of clusters with different diffusion coefficients [18, 19]. Such form of molecular aggregation (clustering) especially at higher vapour concentrations could modify sorption and diffusion behaviour of polymer membranes and therefore affect their transport properties [19-21].

Other complications occur in semi-crystalline polymers, in which the amorphous phase may be either rubbery or glassy. Since the crystalline phase is generally considered to be impermeable, with increasing crystallinity the amount of permeable amorphous phase decreases while simultaneously the diffusion path becomes increasingly tortuous. The latter depends furthermore on the crystal size, shape and orientation. Finally, confinement of the amorphous phase in small domains between the crystallites usually reduces the polymer mobility and increases the T_g of the amorphous phase, with possible consequences for the diffusion coefficient. Therefore accurate description of diffusion through semi-crystalline polymers is considerably more complex than description of diffusion through glassy or rubbery polymers [1, 3, 22-24].

2.1. Permeability of Polymer Membranes

The permeability coefficient P is defined as the amount of the permeant (in m^3 or mol) penetrating at the standard conditions (STP) (the temperature

273.15 K and the pressure 101.325 kPa) through the unit area of the membrane during unit time interval and under unit pressure gradient. Its dimension is (mol m m^{-2} s^{-1} Pa^{-1}) or (m^3(STP) m m^{-2} s^{-1} Pa^{-1}) or (m^2 s^{-1} Pa^{-1}). In American and English literature very often the permeability coefficient has dimension (cm^3(STP) cm cm^{-2} s^{-1} (cm $Hg)^{-1}$), because Barrer defined P as diffusion flow of gas (vapours) in cm^3 s^{-1} (measured at standard conditions) through the membrane area 1 cm^2, when Δp is expressed in cm Hg and membrane thickness is 1cm. [1]

$$1\ \text{Barrer} = 10^{-10}\ \frac{\text{cm}^3(\text{STP})\text{cm}}{\text{cm}^2\ \text{s}\ (\text{cm Hg})} = 7{,}5.10^{-18}\ \frac{\text{m}^3(\text{STP})\ \text{m}}{\text{m}^2\ \text{s Pa}} = 3.346.10^{-16}\ \frac{\text{mol m}}{\text{m}^2\ \text{s Pa}}$$

Vapour permeation through nonporous polymer membrane can be divided into two steps: sorption into the polymer and subsequent diffusion through it. If the sorption of gas (vapours) in the polymer membrane is very low, then according to Henry's law [1], the concentration of gas (vapours) inside the membrane (c) is proportional to the applied pressure (p) through the solubility (sorption) coefficient, S, as

$$c = S\,p \qquad [T] \tag{2}$$

According to the Fick's first law for one-way diffusion (along the x-axis) at constant temperature and pressure:

$$J = -D(c)\frac{\partial c}{\partial x} \qquad [T,\ p] \tag{3}$$

where J is the density of molar diffusion flow, $D(c)$ is concentration-dependent diffusion coefficient, $\partial c\ /\ \partial x$ is a gradient of molar concentration of a substance in a membrane.

Differentiation of equation (2) at constant S and substitution in equation (3) gives the following equation

$$J = -P\frac{\partial p}{\partial x} \tag{4}$$

where $\partial p\ /\ \partial x$ is a gradient of pressure - the driving force of permeability. After integration of equation (4) at initial and boundary conditions ($p = p_1$ at x

= 0 and $p = p_2$ at $x = l$) in time $\tau \geq 0$ and rearrangement the following relation for P is obtained

$$P = \frac{J_S\, l}{p_1 - p_2} \tag{5}$$

where J_S is the density of diffusion flow at steady-state, l is the membrane thickness, p_1 and p_2 are the constant, equilibrium pressures of vapours at the both membrane sides ($p_1 > p_2$).

Experimentally it was proved that in real systems polymer+vapour(s) the diffusion coefficient and also the solubility one are not constant. Nonideality expresses itself mainly in nonlinear increase of vapour solubility with pressure [1, 22, 23, 25]. Therefore, the permeability coefficient P of vapours in polymers is function of vapour pressure $P = P(p)$ and the mean value of (integral) permeability coefficient is defined as

$$\overline{P} = \frac{1}{p_1 - p_2} \int_{p_2}^{p_1} P(p)\,\mathrm{d}p \tag{6}$$

Further, applying the stationary diffusion theory [26] on the experimental vapour sorption data in flat polymer membranes makes it possible to estimate the permeability coefficients from equilibrium sorption data. The procedure is based on assumption of establishing a hypothetical steady state between the surface and the center of the membrane (half of membrane thickness $l / 2$), where the flux J of component i and the concentration gradient $\partial c_i / \partial x$ from the 1st Fick´s law can be replaced by their time-period averages, as

$$< J_i > = \overline{D} \left\langle \frac{\partial c_i}{\partial x} \right\rangle, \tag{7}$$

where $\overline{D}$ is the integral diffusion coefficient which in narrow concentration interval represents the mean value of the differential diffusion coefficient [22, 25, 26]. On experimentally determined sorption kinetics data could be applied kinetic equation of the first order due to the difference of sorbate activities inside and outside of the polymer membrane. The relationship for sorbed mass of vapour m in time t is given by equation

$$m(t) = m_{\infty}\left[1 - \exp\left(-\frac{t}{C}\right)\right], \tag{8}$$

where m_{∞} is the mass uptake at infinite time (in sorption equilibrium), C is a velocity constant of sorption experiment (obtained by data fitting), and J is the density of mass flux, which is defined as

$$J = \frac{1}{2A}\left(\frac{\partial m}{\partial t}\right) \tag{9}$$

where A is a surface area on one side of the membrane (i.e. one half of membrane thickness).

Equation (9) can be expressed according to equation (8) in form

$$J = \frac{m_{\infty}}{2CA}\exp\left(-\frac{t}{C}\right). \tag{10}$$

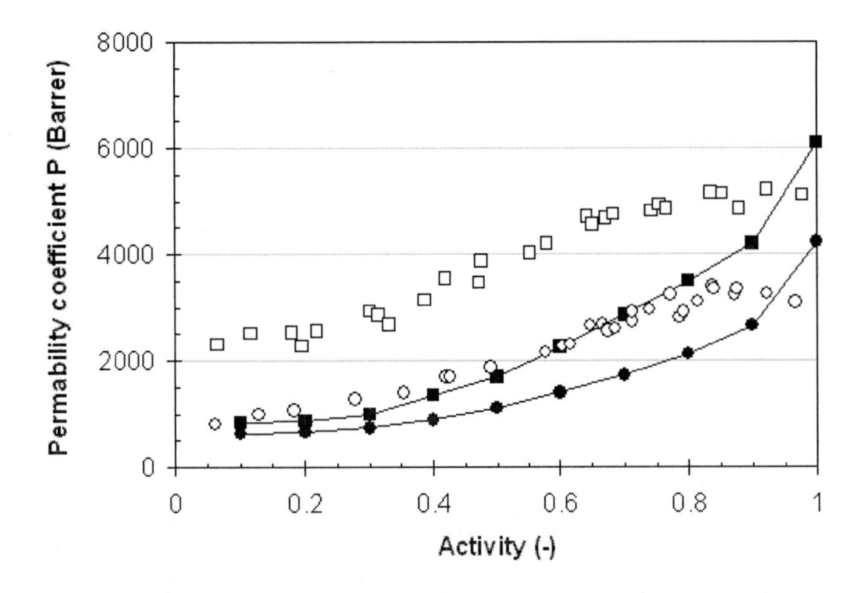

Figure 2. Comparison of the toluene (squares) and heptane (dots) permeability coefficients in low density polyethylene at 298.15K and at atmospheric pressure or at the appropriate vapour pressure from permeation measurements (full symbols) and estimated values from sorption experiments (empty symbols)[21].

The time-period average value of flux density can be obtained by integration of equation (10):

$$< J > = \frac{1}{m_\infty 2CA} \int_{m=0}^{m=m_\infty} (m_\infty - m)\, dm = \frac{m_\infty}{4CA} \tag{11}$$

Finally, estimated value of permeability coefficient (in units: m mol m^{-2} s^{-1} Pa^{-1}) is (with respect to total membrane thickness) given by equation

$$P = \frac{m_\infty\, l}{8 C\, A M\, p}\,, \tag{12}$$

where M is the molar mass of the sorbed vapour, l is the membrane thickness and p is the equilibrium vapour pressure.

The values of the permeability coefficient obtained by different techniques are usually not completely the same because the result often depends on the experimental conditions, which often change from method to method, and on the sensitivity of the method itself [21, 26]. An example is given in Figure 2.

2.2. Diffusion in Polymer Membranes

The diffusion coefficient D is defined as the amount of vapours in mol (at standard conditions) which at constant pressure and temperature penetrates through unit surface per 1 second under unit concentration gradient. Therefore the dimension of D is m^2 s^{-1} [1].

As described above, according to Fick's first law, the density of diffusion flow is directly proportional to a concentration gradient of the diffusant and diffusion proceeds in the direction of lower concentration.

The density of diffusion flow indicates the amount of substance (in mol) which penetrates through the unit of membrane area per unit of time

$$J = \frac{1}{A} \frac{\mathrm{d}n}{\mathrm{d}\tau} \tag{13}$$

where J is density of molar diffusion flow, n is amount of substance, A is membrane area, τ is time. The dependence of density of diffusion flow on time at non-stationary diffusion formulates Fick´s second law [22, 25], which can be derived from Fick´s 1st law by the balance of diffusant in a volume element of the system. The time change of diffusant concentration in volume element dV equals to the difference of amount of diffusant which enters this volume element and the diffusant which leaves the volume element:

$$\frac{\partial c}{\partial \tau} = \frac{A \cdot \left[J|_x - J|_{x+\mathrm{d}x} \right]}{\mathrm{d}V} = \frac{A \cdot \left[J|_x - J|_{x+\mathrm{d}x} \right]}{A\mathrm{d}x} = \frac{1}{\mathrm{d}x}\left[J|_x - J|_{x+\mathrm{d}x} \right] \quad (14)$$

The density of diffusion flow in point x + dx can be expressed by a Taylor series. When members with second and higher derivation are neglected:

$$J|_{x+\mathrm{d}x} = J|_x + \frac{\partial J}{\partial x}\mathrm{d}x \quad (15)$$

Substitution of this series in equation (14) leads to:

$$\frac{\partial c}{\partial \tau} = -\frac{\partial J}{\partial x} \quad (16)$$

Substitution of Fick´s first law in equation (16) yields the Fick´s second law for nonstationary one-way diffusion without chemical reaction

$$\frac{\partial c}{\partial \tau} = \frac{\partial}{\partial x}\left[D(c)\frac{\partial c}{\partial x} \right] \qquad [T, p] \quad (17)$$

For a constant diffusion coefficient we obtain:

$$\frac{\partial c}{\partial \tau} = D\frac{\partial^2 c}{\partial x^2} \qquad [T, p] \quad (18)$$

If the diffusion coefficient depends on the concentration, which at nonstationary diffusion is a function of position, then equation (17) changes to:

$$\frac{\partial c}{\partial \tau} = \frac{\partial D(c)}{\partial x} \cdot \frac{\partial c}{\partial x} + D(c) \cdot \frac{\partial^2 c}{\partial x^2} \qquad [T, p] \tag{19}$$

The dependence of the diffusion coefficient on the position is given by relation

$$\frac{\partial D}{\partial x} = \frac{\partial D}{\partial c} \cdot \frac{\partial c}{\partial x} \tag{20}$$

By combination of equations (19) and (20) Fick´s law for concentration-dependent diffusion is obtained

$$\frac{\partial c}{\partial \tau} = \frac{\partial D(c)}{\partial c} \cdot \left(\frac{\partial c}{\partial x}\right)^2 + D(c) \cdot \frac{\partial^2 c}{\partial x^2} \qquad [T, p] \tag{21}$$

If the concentration change is small, the value of $\partial D / \partial c$ is also small in comparison with the value of D. The second term on the right side of equation (21) can then be neglected and instead of a differential diffusion coefficient one can use a constant mean value. In real systems the density of diffusion flow is proportional to the gradient of chemical potential of a diffusant and therefore all derived relations are valid for ideal polymer-gas, polymer-vapour or polymer-liquid systems only [1, 22, 23].

For diffusion of vapours through a membrane with thickness l and constant surface concentration of diffusants c_1 and c_2, ($c_1 > c_2$), then at steady state, when the concentration of the diffusant does not change with time [$\partial c / \partial \tau = 0$], equation (17) becomes:

$$\frac{\partial}{\partial x}\left[D(c)\frac{\partial c}{\partial x}\right] = 0 \qquad [T, p] \tag{22}$$

After double integration of equation (22) at initial and boundary condition ($c = c_1$ at $x = 0$ and $c = c_2$ at $x = l$, $\tau \geq 0$) and combination with equation (3), for the density of stationary diffusion flow J_S can be obtained

$$J_S = -\frac{1}{l}\int_{c_1}^{c_2} D(c)\,\mathrm{d}c = \bar{D}\left(\frac{c_1 - c_2}{l}\right) \qquad [T, p] \tag{23}$$

where the integral diffusion coefficient $\bar{D}$ is defined as

$$\bar{D} = \frac{1}{c_1 - c_2}\int_{c_2}^{c_1} D(c)\,\mathrm{d}c \tag{24}$$

and it presents the mean value of diffusion coefficient.

As follows from equation (23), the density of stationary diffusion flow is directly proportional to the concentration difference ($c_1 - c_2$) and inversely proportional to the membrane thickness l.

When the concentration c_2 is equal to zero, then the integral diffusion coefficient reduces to:

$$\bar{D} = \frac{1}{c_1}\int_{0}^{c_1} D(c)\,\mathrm{d}c \tag{25}$$

From equation (25) it is evident, that at $c_2 = 0$ the differential diffusion coefficient $D(c)$ is related to the integral diffusion coefficient $\overline{D}$ by relation

$$D_{c=c_1} = \overline{D} + c_1\frac{\mathrm{d}\overline{D}}{\mathrm{d}c_1} \tag{26}$$

At constant diffusion coefficient $\overline{D} = D$ the diffusant concentration inside the membrane is thus a linear function of a coordinate x

$$c = c_1 + \left(\frac{c_2 - c_1}{l}\right)x \tag{27}$$

and analogously as in equation (23) it follows that

$$J_S = D\left(\frac{c_1 - c_2}{l}\right) \qquad [T, p] \tag{28}$$

Generally, the diffusion coefficient is a function of temperature, pressure and concentration c. In anisotropic materials it is also a function of position x. If the diffusion coefficient is also a function of time, then the Fick´s laws can not be applied and diffusion is non-fickian [1, 22, 23].

2.3. Sorption in Polymer Membranes

The solubility (sorption) coefficient S usually is defined as the amount of vapours in m^3 (at standard conditions) which is dissolved in 1 m^3 of solvent (polymer) at partial pressure 1 Pa and at the given temperature. Therefore the dimension of S is $m^3_{STP}\ m^{-3} Pa^{-1}$. The solubility coefficient can be defined also as the amount of vapours in grams, which is dissolved in 1 gram of polymer at partial pressure of 1 Pa ($g\ g^{-1}\ Pa^{-1}$) [1].

Depending on the system, sorption of vapours in polymers is not only governed by linear Henry's law [equation (2)], but also the other types of sorption isotherms can occur (Langmuir's isotherm or by Flory-Huggins's isotherm) [1, 8] (see Figure 3.). The sorption isotherms express the dependence of the equilibrium concentration of sorbate, c, in a polymer on the applied pressure, p, at a given temperature.

For polymers in their glassy state, the sorption has mostly the dual mode sorption character [1, 8, 10, 24], in which the total concentration of vapours in the membrane is the sum of two contributions

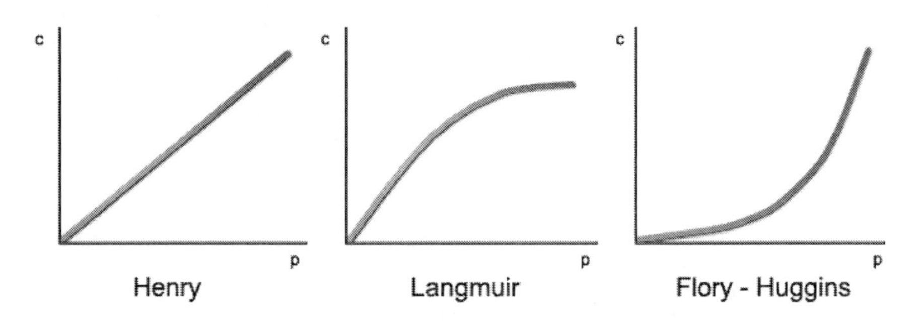

Figure 3. Schematic representation of different types of sorption isotherms.

$$c = c_1 + c_2 \tag{29}$$

The first term in equation (29) is identical to equation (2) and corresponds to the solution (sorption) of a compound in a membrane according to Henry's law, while the second term, corresponding with Langmuir isotherm, represents the sorption of a compound in membrane micro-pores, voids or defects

$$c_2 = a \frac{b\,p}{1 + b\,p} \tag{30}$$

where p is the gas or vapour pressure and a, b are constants.

Equations (29) and (30) give together

$$c = S\,p + a\,\frac{b\,p}{1 + b\,p} \tag{31}$$

It can be proved that during the dual sorption at high pressures ($bp >> 1$) the density of gas (vapours) diffusion flow follows equation (2) as well. At the low pressures ($bp << 1$) the density of gas (vapours) diffusion flow in the steady-state is given by equation

$$J = -D_{ef} S\,\frac{\partial p}{\partial x} = -P\frac{\partial p}{\partial x} \tag{32}$$

It this case the permeability coefficient P is the product of the effective diffusion coefficient D_{ef} and the solubility (sorption) coefficient S. The effective diffusion coefficient is defined as

$$D_{ef} = \left(\frac{D\,S}{S + a\,b} \right) \tag{33}$$

For the integral solubility coefficient, which is also function of the pressure we obtain

$$\bar{S} = \frac{1}{p_1 - p_2} \int_{p_2}^{p_1} S(p)\,\mathrm{d}p \tag{34}$$

2.4. Dependence of Transport Parameters on Temperature

The temperature dependence of P, D, and S coefficients is usually expressed by an Arrhenius type of equation [1]. Therefore the dependence of permeability coefficient P on the temperature is given by equation

$$P = P_0 \exp\left(-\frac{E_p}{RT}\right) \tag{35}$$

where P_0 is a temperature-independent and constant pre-exponential factor, E_p is the activation energy of permeation, R is the universal gas constant and T is the absolute temperature (K).

Analogously, the dependence of the diffusion coefficient on temperature is described by a similar equation

$$D = D_0 \exp\left(-\frac{E_d}{RT}\right) \tag{36}$$

where D_0 is a constant pre-exponential factor and E_d is the activation energy of diffusion.

The same dependence of solubility (sorption) coefficient on the temperature is expressed by the equation

$$S = S_0 \exp\left(-\frac{\Delta H_s}{RT}\right) \tag{37}$$

where S_0 is also a pre-exponential factor and ΔH_s is an enthalpy of solution, which can have a positive or negative value.

Activation energies of permeability and diffusion have normally positive values and therefore the coefficients of permeability and diffusion generally increase with increasing temperature. The value of solubility (sorption) coefficient increases with temperature only in the case when the enthalpy of solution is positive.

2.5. Correlation of Transport Parameters with Thermodynamic Critical Quantities

Diffusion is an activated process which takes place by a 'hopping' mechanism of the penetrant molecule from one free volume element in the polymer to the next element [2]. The frequency of such events depends of course on the mobility of the polymer matrix, i.e. on the glassy or rubbery state of the polymer, and on the dimensions of the penetrant molecule. For this reason the diffusion coefficient is often related to the critical volume of the penetrant, as a representative measure of its size [4].

$$D = \frac{\gamma}{V_c^{\eta}} \tag{38}$$

where γ and η are constants. An example of the dependence of D on V_c is shown in Figure 4. This trend is not completely universal, because it clearly depends also on the chemical nature of the penetrant species, as can be seen form the different position of the alcohols and water in the graph compared to the hydrocarbons or other equivalents.

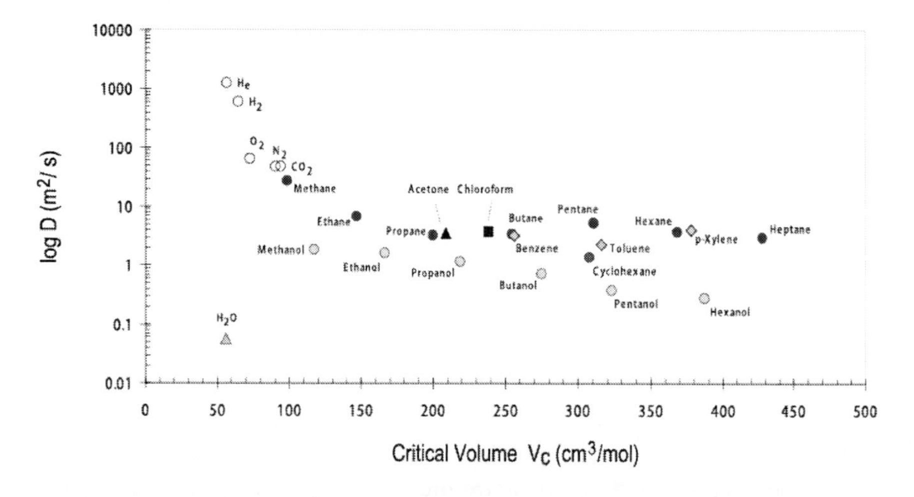

Figure 4. Correlation of the gas and vapour integral diffusion coefficients in low density polyethylene at 298.15K and at atmospheric pressure or at the appropriate vapour pressure with critical volume of the diffusant [this work, 21, 26-28].

Analogously, the penetrant solubility in the polymer matrix depends on its condensability at the given conditions. Often the critical temperature, T_c, of the penetrant is used as a measure for its condensability and it was found that the logarithm of the solubility can be correlated with the square of T_c [3, 4]

$$\log S = N + M T_c^2 \quad (39)$$

or

$$\log S = N + M\left(\frac{T_c}{T}\right)^2 \quad (40)$$

in which M and N are constants. This trend is illustrated in Figure 5.

The dependence of the diffusion coefficient on the molecular size is much stronger than the dependence of the solubility on its critical temperature. Therefore, the permeability coefficient, the product of D and S, is also generally correlated with the critical volume of the penetrant, i.e. with its size, according to the following equation with two constants a, b:

$$P = a \exp V_c^{\,b} \quad (41)$$

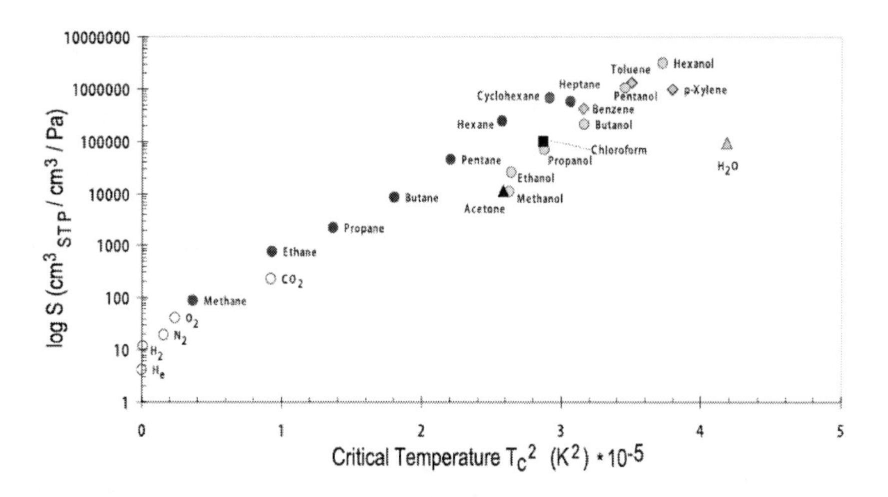

Figure 5. Correlation of the gas and vapour integral sorption coefficients in low density polyethylene at 298.15K and atmospheric pressure or at appropriate vapour pressure with square of critical temperature of sorbate [this work, 21, 26-28].

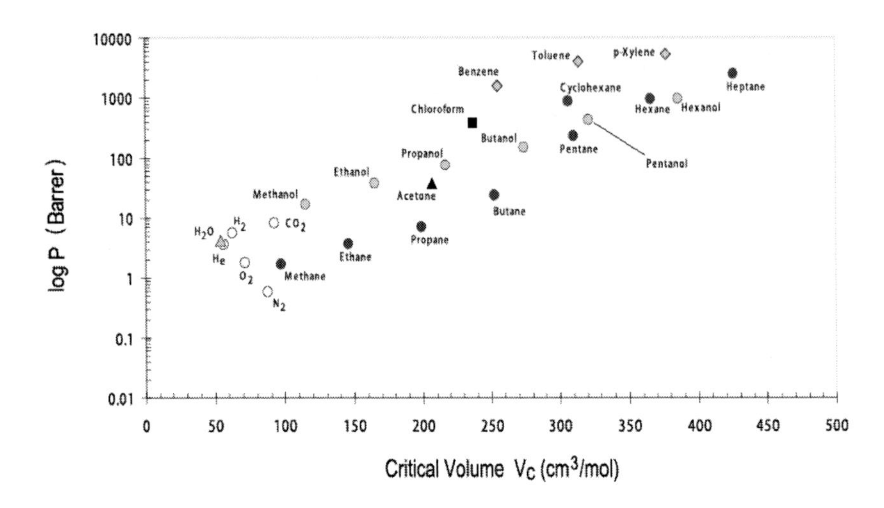

Figure 6. Correlation of the gas and vapour permeation coefficients in low density polyethylene at 298.15K and atmospheric pressure with critical volume of the penetrant [this work, 21, 26-28].

The result is shown in Figure 6. Also in this case there is a significant difference in the trend of the alkanes and the related alcohols, indicating that other molecular properties are also important.

3. Experimental Methods for Determination of Vapour Transport in Polymers

3.1. Time-Lag Permeation

One of the most common experimental methods to obtain information on the diffusivity of gaseous species or vapours in dense polymeric materials is by time lag measurements. This method is based on the penetration theory and on the very characteristic trend in the permeation rate of a penetrant through a dense membrane, from the very first moment of exposure until permeation of the penetrant reaches the stationary state. This is schematically displayed in Figure 7. Immediately after the exposition of the membrane to the gas or to the vapour equilibrium between the feed pressure, p_f, and penetrant concentration, c_1, at the interface is formed, in which the equilibrium constant is the gas solubility, S (Eq. 2) {see above}.

In this phase a convex concentration profile is established inside the membrane. Only after a finite time, when the penetrant molecules reach the permeate side of the membrane, the permeate pressure starts rising. This is followed by a transient period, in which the convex concentration profile gradually becomes linear, and if $p_f = p_1 >> p_2 = p_P$ the pressure increase rate of permeate becomes constant.

The practical side of this method is usually as follows: the membrane, placed in a permeation cell, is first carefully evacuated and then suddenly exposed to the penetrant species at one side, while simultaneously the penetrant concentration is measured at the other side.

The most straightforward way to do this is to measure the pressure of a closed permeate chamber. Various other methods are possible as well.

Mathematically this can be described as follows: If a penetrant-free membrane is exposed to the penetrant at the feed side at $t = 0$ and the penetrant concentration is kept very low at the permeate side, then the total amount of penetrant, Q_t, passing through the membrane in time t is given by [1]:

$$\frac{Q_t}{l \cdot c_1} = \frac{D \cdot t}{l^2} - \frac{1}{6} - \frac{2}{\pi^2} \sum_1^{\infty} \frac{(-1)^n}{n^2} \exp\left(-\frac{D \cdot n^2 \cdot \pi^2 \cdot t}{l^2} \right) \tag{42}$$

in which c_1 is the penetrant concentration at the membrane interface at the feed side, l is the membrane thickness [m] and D is the diffusion coefficient [m^2/s]. For a setup with a constant permeate volume in which the pressure, p_t, is measured as a function of time, Eq. (42) becomes:

$$p_t = \frac{RT \cdot A \cdot l}{V_P \cdot V_m} \cdot p_f \cdot S \left(\frac{D \cdot t}{l^2} - \frac{1}{6} - \frac{2}{\pi^2} \sum_1^{\infty} \frac{(-1)^n}{n^2} \exp\left(-\frac{D \cdot n^2 \cdot \pi^2 \cdot t}{l^2} \right) \right) \tag{43}$$

In which p_t is the permeate [bar] pressure at time t [s], R is the universal gas constant [$8.314 \cdot 10^{-5}$ m^{3}bar/(mol·K)], T is the absolute temperature [K], A is the exposed membrane area [m^2], V_P is the permeate volume [m^3], V_m is the molar volume of a gas (vapours) at standard temperature and pressure [$22.41 \cdot 10^{-3}$ m$^3_{STP}$/mol at 0°C and 1 atm], p_f is the feed pressure [bar] and S is the gas (vapours) solubility [m$^3_{STP}$/(m^3 bar)].

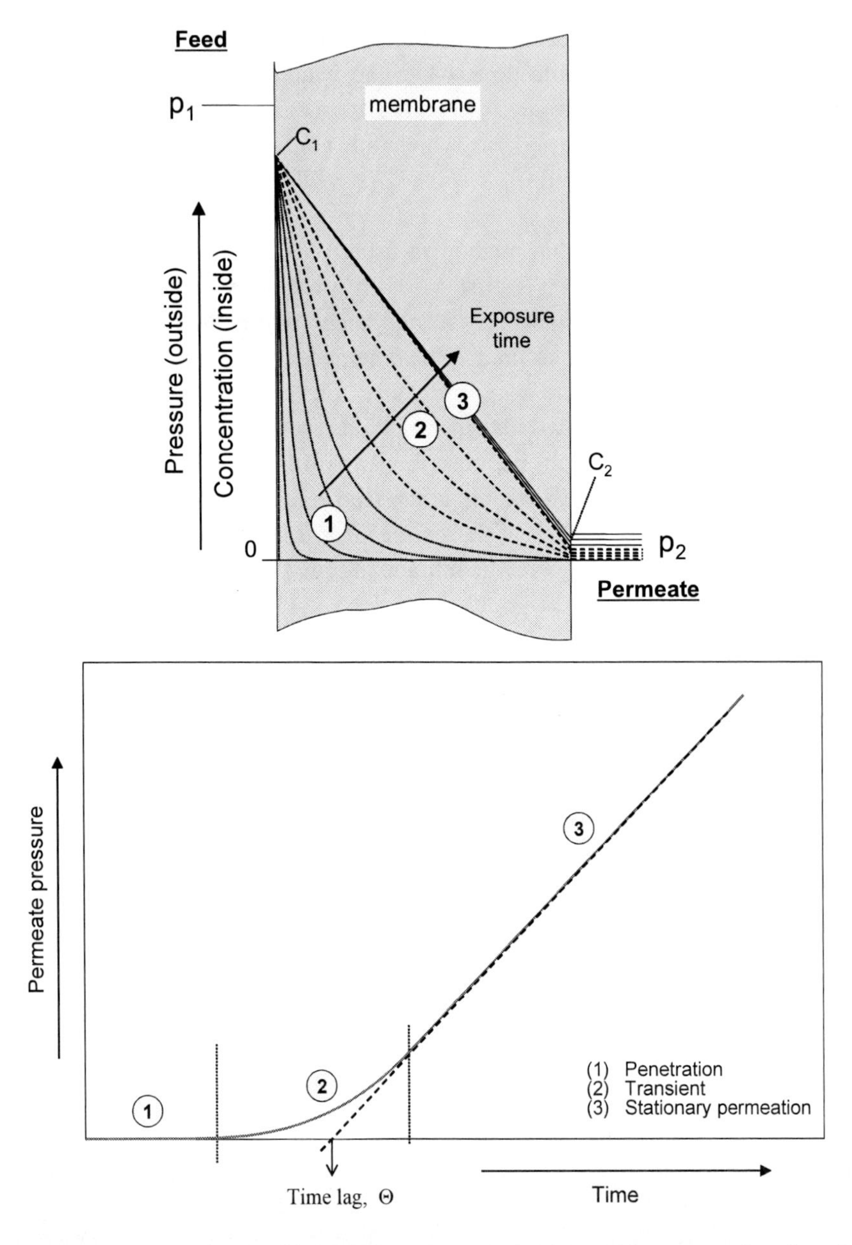

Figure 7. Concentration profile of the penetrant across the membrane as a function of time (left) and permeate pressure (right). Stage 1: penetration, stage 2: transient, stage 3: stationary permeation. The intercept of the tangent to the stationary pressure increase curve with the time axis indicates the so called time lag.

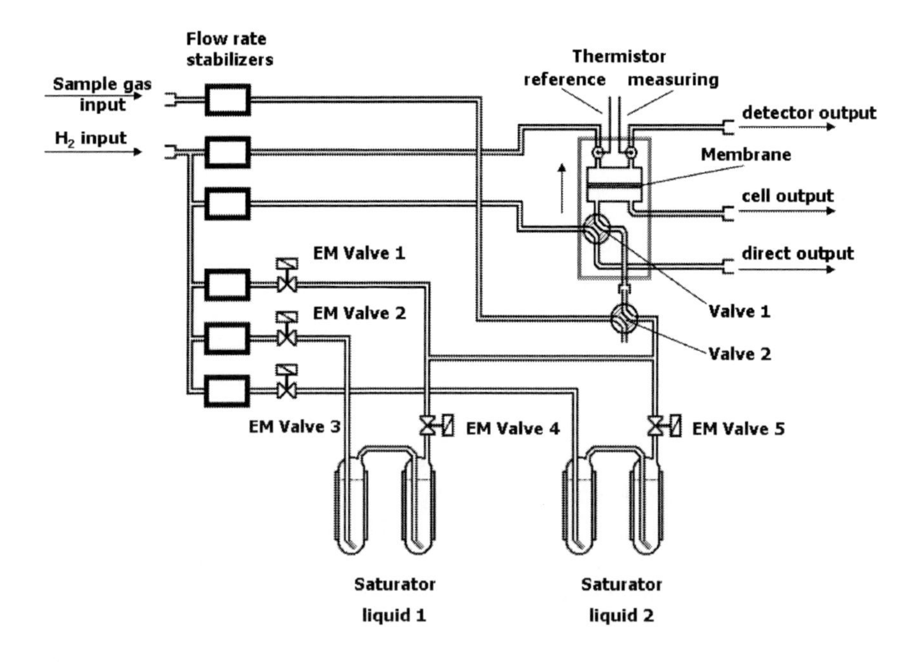

Figure 8. The schematic drawing of the differential permeation apparatus [21, 26].

At long times the exponential term approaches to zero and Eq. (43) reduces to:

$$p_t = \frac{RT \cdot A \cdot l}{V_P \cdot V_m} \cdot p_f \cdot S\left(\frac{D \cdot t}{l^2} - \frac{1}{6}\right) = \frac{RT \cdot A}{V_P \cdot V_m} \cdot \frac{p_f \cdot S \cdot D}{l}\left(t - \frac{l^2}{6D}\right) \quad (44)$$

Thus, at long times a plot of p_t versus time describes a straight line which, upon extrapolation, intersects the time axis at $t = l^2 / 6D$, defined as the time lag, Θ [s].

$$\Theta = \frac{l^2}{6D} \quad (45)$$

With this equation the diffusion coefficient can be obtained by time lag measurements if the membrane thickness is known.

3.2. Differential Permeation Method

The experimental apparatus working on the basis of the differential permeation method [21] contains a temperature-controlled permeation cell, partitioned by a flat polymer membrane with known thickness and area into two identical compartments (Figure 8). In the feed-side compartment the membrane is in contact with a constant carrier gas (hydrogen or helium) plus vapour flux and in the permeate-side compartment with pure carrier gas flux. The pressure on both membrane sides must be identical. Changes of thermal conductivity in time on the permeate side with respect to the value of pure carrier gas are determined by a pair of thermal conductivity detectors (thermistors) built into a Wheatstone resistance bridge. This enables the evaluation of D from the permeation transient, and P from the steady-state. Vapour permeation measurements can be carried out at different values of feed carrier gas saturations, ranging up to 100 % with respect to each tested vapour pressure at given temperature (100 % saturation means that the vapour partial pressure in the carrier gas stream is equal to the value of the vapour pressure at that given temperature).

The permeability coefficients (in [m^3(STP) m] / [m^2 s Pa]) of the measured vapours [21, 22, 25] in the appropriate concentration interval are calculated by:

$$P = \frac{k\,N_{ST}\,v\,l}{p_{\mathrm{V}}}\,\frac{T_0\,p_{\mathrm{B}}}{A\,T\,p_0}, \tag{46}$$

where k is an apparatus constant (determined by calibration), N_{ST} is the voltage signal in steady-state, l is the membrane thickness, v the flow rate of the mixture, A is the membrane area, T is the temperature of the experiment, T_0 is the standard temperature 273.15 K, p_0 is the standard pressure 101.325 kPa, p_{V} is the partial pressure of the vapours in the carrier gas (for gases $p_{\mathrm{V}} = p_{\mathrm{B}}$) and p_{B} is the barometric pressure. The effective (integral) diffusion coefficients were obtained from experimental data fitting by equation

$$J(t) = \left(\frac{D\,c_1}{l}\right)\left[1 + 2\sum_{n=1}^{\infty}(-1)^n \exp\left(-\frac{D\,n^2\,\pi^2\,t}{l^2}\right)\right] \tag{47}$$

which was derived [18, 33] from second Fick's law under appropriate initial and boundary conditions and where the flux $J(t)$ through the membrane is measured as a function of time by a thermal-conductivity detector.

3.3. Gravimetric Sorption Method

The vapour sorption experiments can be performed by several methods [1]. One of them is utilizing a quartz (McBain's) spiral balance [21, 26] (Figure 9). The polymer membrane is appended on a calibrated quartz spiral balance in a carefully evacuated glass tube and then suddenly exposed to the sorbate species. The elongation of the spiral due to vapour sorption into the polymer membrane, and thus to a weight change of the sample, is monitored by an optical system from the beginning of the measurement until reaching the equilibrium state.

In case of the vapours sorption experiments in thin polymer film, a similar phenomenon to permeation (see Figure 7) takes place but then in a symmetrical way [21]. The difference is that in the beginning of sorption experiment is high vapour concentration at both membrane sides and vapour molecules started to penetrate into the film from both sides simultaneously. This initial stage is important for kinetics analysis because the transport parameters, usually evaluated on the basis of the initial phase of the experiment, are mainly determined by the lowest resistance in the membrane in the most plasticized zone near the surface. Therefore, differences in the shape of the concentration profiles in permeation and sorption experiments are reflected in the values of the integral diffusion coefficient [21].

The sorption coefficient S is determined from equilibrium sorption data while the integral (effective) diffusion coefficients D can be evaluated from sorption kinetics by fitting of experimental data by equation (48) derived from Fick's second law under appropriate initial and boundary conditions [22, 25].

$$\frac{m_t}{m_\infty} = 1 - \frac{8}{\pi^2} \sum_{i=0}^{\infty} \frac{1}{(2i+1)^2} \exp\left[-\frac{(2i+1)^2 \pi^2 D t}{l^2} \right] \tag{48}$$

where m_t is the mass uptake at time t and m_∞ is the mass uptake at time of reaching the sorption equilibrium and l is the membrane thickness.

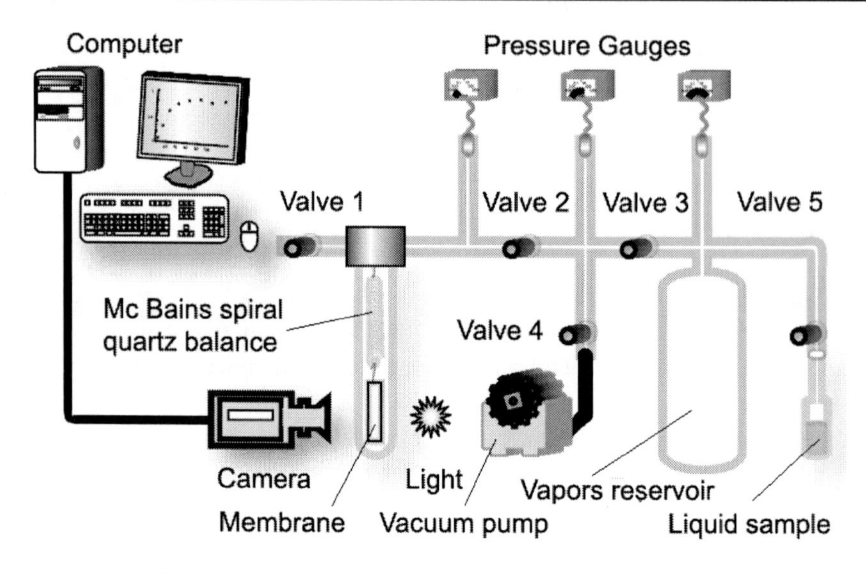

Figure 9. The schematic drawing of the gravimetric sorption apparatus [21, 26].

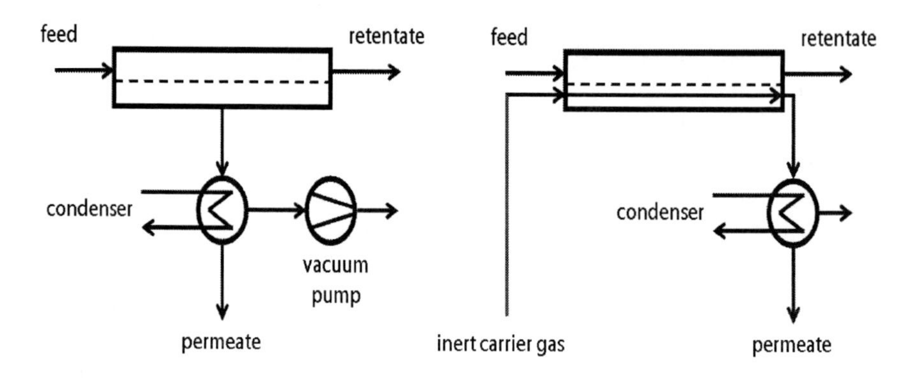

Figure 10. Schematic diagram of the pervaporation process with a downstream vacuum pump (left) or a downstream carrier gas (right) [29].

3.4. Pervaporation

Pervaporation is a separation process in which minor components of a liquid mixture are preferentially transported through a non-porous perm-selective membrane and collected by partial vaporization and subsequent condensation. During the pervaporation process one side of the membrane is in direct contact with liquid phase and permeate is removed from the other side of the membrane in vapour phase by a vacuum (vacuum pervaporation) or by a sweep gas (sweep gas pervaporation) (Figure 10). The driving for of the separation process is the chemical potential gradient, which is a function of pressure and concentration [1, 29-36].

Pervaporation is often seen as an emerging technology in environment clean-up operations, especially in the removal of volatile organic compounds (VOCs) from industrial wastewaters or from contaminated ground waters [30-36]. To get higher productivity of pervaporation it is necessary to use this technique as a continuous process and to use spiral wound or hollow fiber modules with a relatively high membrane surface to volume ratio.

The municipal waste, traffic, industrial and agricultural operations are producing more and more VOCs like trichloroethylene, tetrachloroethylene and petroleum-based solvents (benzene, toluene, ethyl benzene, and xylenes (BTEX)) which are contaminating the ground water. Their low solubility in water hinders utilization of classical separation technologies such as distillation because the amount of VOCs dissolved in water is too small to be economically removed. The air stripping and activated carbon sorption were deployed for the task, however, the former is sensitive to fouling and merely turns a water pollution problem into an air pollution issue while the latter needs costly regeneration steps and may not be suitable for VOCs that are easily displaced by other organic compounds. In the last two decades, a growing effort has been devoted to feasibility studies on the pervaporation process for VOCs removal from water in order to determine if this technology is technically and economically feasible for this application. Indeed, pervaporation seams to be a promising technology in treating dilute VOCs in either ground water or aqueous effluents [30-36].

In the case of one-dimensional permeation across a membrane, combining of the permeation flux of each permeating component (Eq. 3) with a concentration-dependent diffusion coefficient in the form [1]

$$D(c) = D_0 \exp(kc) \tag{49}$$

where: k is a constant, D_0 is the diffusion coefficient if the concentration of permeating component is $c_0 = 0$ provides a relation for a concentration dependent permeation flux:

$$J = -D_0 \exp(kc)\left(\frac{\partial c}{\partial x}\right) \tag{50}$$

If we integrate eq. (50) with initial and boundary condition: $c = c_0$, if $x = 0$ and $c = 0$, if $x = l$:

$$\int_0^l J dx = -D_0 \int_{c_0}^{0} \exp(kc)\, dc \tag{51}$$

then we obtain a relation of permeation flux with concentration dependent diffusion coefficient:

$$J = \frac{D_0}{kl}\left[\exp(kc_0) - 1\right]. \tag{52}$$

The selectivity of pervaporation separation process is usually represented by the separation factor, α of the permeating components (i and j) of a binary liquid mixture:

$$\alpha_{ij} = \frac{\left(c_i/c_j\right)^p}{\left(c_i/c_j\right)^f} \tag{53}$$

where p means permeate phase and f feed phase. Sometimes, also, the enrichment factor of preferential permeating components is used, β_i:

$$\beta_i = \frac{\left(c_i\right)^p}{\left(c_i\right)^f} \tag{54}$$

In very dilute systems, the concentration of the component j (solvent) in Eqs. (53) and (54) will approach value equal to 1 in both the feed and the permeate. The separation factor will therefore be close to the value of the enrichment factor, $\alpha_{ij} \approx \beta_i$.

CONCLUSION

The mass transport of VOCs in and through polymers is in most cases accompanied by vapour concentration-evoked effects (swelling, plasticization, chain flexibility, reorganisation of dynamic free volume elements etc.) with a great impact on polymer structure and its behaviour. Multi-component VOCs mass transport in polymers, in term of its description by Fick's laws, irreversible thermodynamics, generalised Maxwell-Stefan approach (GMS) or models combining GMS or SDM with lattice-based theories etc., constitutes a rather complicated problem because the concentration dependent solubility or diffusivity of one component in the mixture can be influenced significantly by the other component(s). In real systems (VOCs in air or in water), such mutual effects could contribute positively or negatively to the total mass transport of components and therefore they can play an essential role in designing and tailoring of appropriate membrane separation processes.

ACKNOWLEDGMENTS

The authors are thankful for financial support of the Ministry of Education, Sports and Youth MSM (Grant No. 6046137307) and Grant Agency of Czech Republic GACR (Grant No. 106/10/1194).

REFERENCES

[1] Mulder, M. *Basic Principles of Membrane Technology;* Kluwer Academic Publishers: Dordrecht, NL, 1998.

[2] Baker, R. W. *Membrane Technology and Applications;* John Wiley & Sons Ltd: Chichester, UK, 2004.

[3] Matteucci, S.; Yampolskii, Y.; Freeman B. D.; Pinnau I. In Materials Science of Membranes for gas and vapor separation; Yampolskii, Y.; Pinnau, I.; Freeman, B. D.; Eds.; *Transport of Gases and Vapors in Glassy and Rubbery Polymers* (Chapter 1); John Wiley & Sons: Chichester, UK, 2006; 1-48.

[4] Yampolskii, Y.; Freeman, B. D.; Eds.; *Membranes gas separation;* John Wiley & Sons: Chichester, UK, 2010.

[5] Wijmans, J.G.; Baker, R.W. in Book *Materials Science of Membranes for gas and vapor separation; Yampolskii,* Y.; Pinnau, I.; Freeman, B. D. ; Eds.; The solution-diffusion model: a unified approach (Chapter 5); John Wiley & Sons: Chichester, UK, 2006; 159-189.
[6] DiBenedetto, A.T.; Paul D.R. J Polym Sci Part A 1964, 2, 1001-1015.
[7] Brandt, W.W. *J Phys Chem* 1959, 63, 1080-1085.
[8] Vieth, W.R. *Diffusion in and through polymers*; Hanser Publishers: Munich, DE, 1991; 15-19.
[9] Kesting, R.E. *Synthetic polymeric membranes – A structural perspective;* John Wiley & Sons: New York, US, 1985.
[10] Vieth, W. R.; Sladek, K. *J. J Colloid Sci* 1965, 20, 1014-1033.
[11] Meares, P. *J Polym Sci* 1958, 27, 405-418.
[12] Fujita, H.; Kishimoto, A.; Matsumoto, K. *Tran Faraday Soc* 1960, 56, 424-437.
[13] Vrentas, J.S.; Hou, A.C.; *J Appl Polym Sci* 1987, 36, 1933-1934.
[14] Vieth, W.R.; Felder, R.M.; Huvard, G.S. In Methods of Experimental Physics; Fava, R.; Ed.; *Permeation, diffusion, and sorption of gases and vapors* (16C); Academic Press: New York, US, 1978; 315-377.
[15] Bontoux, L.G.; Soane, D.S.; *J Appl Polym Sci* 1989, 38, 915-922.
[16] Hopfenberg, H.B.; Frisch, H.L. J Polym Sci Part B Poly Lett 1969, 7, 405-409.
[17] Vrentas, J.S.; Duda, J.L.; Huang, W.J. *Macromolecules* 1986, 19, 1718-1724.
[18] Nguyen, Q.T.; Favre, E.; Ping, Z.H.; Néel, J. *J Membr Sci* 1996, 113, 137-150.
[19] Tokarev, A.; Friess, K.; Machková, J.; Šípek, M.; Yampolskii, Y. *J Polym Sci Part B Polym Phys* 2006, 44, 832-844.
[20] Jansen, J.C.; Friess, K.; Drioli, E. *J. Membr. Sci.* 2011, 367, 141-151.
[21] Friess, K.; Jansen, J. C.; Vopička, O.; Randová, A.; Hynek, V.; Šípek, M.; Bartovská, L.; Izák, P.; Dingemans, M.; Dewulf, J.; Van Langenhove, H.; Drioli, E. *J. Membr. Sci.* 2009, 338, 161-174.
[22] Crank J.; Park, G. S. *Diffusion in Polymers;* Academic Press: London, UK, 1968.
[23] Cussler, E. L. Diffusion, *Mass Transfer in Fluid Systems;* Cambridge University Press: Cambridge, UK, 1997.
[24] Strobl, G. *The Physics of Polymers;* Springer-Verlag: Berlin, DE, 2007.
[25] Crank, J. *The Mathematics of Diffusion;* Clarendon Press: Oxford, UK, 1975.

[26] Friess, K.; Šípek, M.; Hynek, V.; Sysel, P.; Bohatá, K.; Izák, P. *J Membr Sci* 2004, 240, 179-185.

[27] Van Krevelen, D. *Properties of Polymers;* Elsevier: Amsterdam, NL, 1990; 535-582.

[28] Izák, P.; Bartovská, L.; Friess, K.; Šípek, M.; Uchytil, P. *J Membr Sci* 2003, 214, 293-309.

[29] Izák, P.; Friess, K.; Šípek, M. In *Handbook of Membrane Research: Properties, Performance and Applications;* Gorley, S. V.; Ed.; Permeation and Pervaporation Taking Advantage of Ionic Liquids (Chapter 12); Nova Science Publishers: New York, US, 2009; 387-402.

[30] Peng, M.; Vane, L. M.; Liu, S. X. *J. Hazard Mat* 2003, 98, 69–90.

[31] Vane, L. M.; Alvarez, F. R. *Separ Purif Technol* 2001, 24, 67–84.

[32] Shah, M. R.; Noble, R. D.; Clough, D. E. *J. Membr. Sci.* 2004, 241, 257–263.

[33] Ohshima, T.; Kogami, Y.; Miyata, T.; Uragami, T. *J. Membr. Sci.* 2005, 260, 156–163.

[34] Xu, J.; Ito, A. *Desalination and Water Treatment* 2010, 17, 135–142.

[35] Park, Y. I.; Yeom, C.K.; Lee, S. H.; Kim, B. S.; Lee, J. M.; Joo, H. J. *J. Ind. Eng.Chem.* 2007, 13, 272-278.

[36] Kim, K.S.; Kwon, T. S.; Yang, J.S.; Yang, J.W. *Desalination* 2007, 205, 87–96.

In: Volatile Organic Compounds
Editors: J. C. Hanks et al. pp. 149-166
ISBN 978-1-61324-156-1

Chapter 5

SOURCES AND ELIMINATION OF VOLATILE ORGANIC COMPOUNDS

Andrés M. Peluso, Horacio J. Thomas and Jorge E. Sambeth[*]
Centro de Investigación y Desarrollo en Ciencias Aplicadas "Dr. Jorge J. Ronco", CINDECA (FCE UNLP – CCT CONICET LA PLATA), 47 N.357 (1900) La Plata, Argentina

ABSTRACT

The volatile organic compounds (VOCs) are defined, according to USEPA, as those organic compounds that, at 20 °C, present a vapor pressure equal or higher than 0.01 KPa. This defines about 200 chemical compounds such alcohols, chlorinated hydrocarbons, esters, etc. excluding CH_4. The European Union defines VOCs as any organic compounds that have a starting boiling point lower or equal to 250 ºC measured in standard atmospheric pressure (101.3 kPa). VOCs are harmful due to different factors; on the one hand, they form part of the photochemical smog and some of them participate in the greenhouse effect and on the other hand, some of them are cancerous or teratogenics. VOCs sources are classified in biogenic, created by nature, and anthropogenic mainly linked to transport and use of solvents. The control of VOCs can be divided in two groups: primary, which is related to technological substitution and secondary, which are related to the

[*] Email: sambeth@quimica.unlp.edu.ar

elimination at the end of pipe. The last technologies are also classified in recovery methods (adsorption, condensation, membranes) and destruction methods (biofiltration, thermal and catalytic oxidation).

INTRODUCTION

The atmospheric contamination is caused by a mixture of substances that come from different origins. The sources can be biogenic such as volcanoes, forest fires, etc. or anthropogenic such as mobile sources, industrial processes, home use of paints and solvents, among others. Among the pollutants found in the air, we can mention CO, NO_x, SO_x, Particulate Material (PM), NH_3 and the volatile organic compounds (VOCs), among others.

According to the American Environmental Protection Agency (USEPA), VOCs are defined as any carbon compound, which forms part of the atmospheric photochemical reactions, except for CO, CO_2, carbides or metallic carbonates, and ammonium carbonate. The European Union defines them as those organic compounds that have a boiling point lower or equal to 250 °C measured at 101.3 KPa. VOCs consist of substances such as alcohols, aromatic hydrocarbons, organic acids, esters, ketones, chlorinated hydrocarbons, etc.

They are generally classified as biogenic and anthropogenic, the latter being 80% of the total emissions. The biogenic sources are classified in: (i) terpenoids, consisting of compounds with C5, C10, C15 and C20 and (ii) other VOCs. Among terpenoids, we can mention isoprene, α-limonene, β-pinene and cumene, and among other biogenic VOCs ethene, butane, acetone and acetaldehyde. It is estimated that the global emissions of biogenic VOCs are 1150 TgC per year which 50% of them are assigned to isoprene.

The anthropogenic VOCs sources are mainly related to the use of fossil fuels in transport, industrial processes and goods manufacturing, calculating that the use of solvents and related products are responsible for around 55% of the total VOCs produced by human beings. Among them, we can find benzene, chloroform, CCl_4, acetone, toluene, propane and percloroethylene.

Besides, there are dissolved VOCs in wastewaters. The main sources are: (a) water provision plants, (b) industries, (c) commercial premises, (d) household water and runoff zones.

In the provision plants, chlorination can give rise to VOCs such as CCl_4, $ClCH_3$ and CH_2Cl_2. Industries such as that of plastics, pesticides, paints, metallurgical among others, release liquid waste with dissolved VOCs such as benzene, phenol, CCl_4, CH_2CHCN, $ClCH_3$, CH_2Cl_2, C_2Cl_4 and C_2HCl_3. Many of these VOCs together with acetone, ethyl acetate, toluene, xylenes and cyclohexanes have been found in wastewaters from houses, small industries and gas stations. VOCs concentration in water is variable since it depends on temperature, partial pressure, water superficial extension, among other factors. However, in 1982, the USEPA analyzed the water from 40 cities finding in the sampling that in 95.8% of the city waters, there was toluene, in 94.8% there was C_2Cl_4, in 91.3% of the cities there was $ClCH_3$ and in 60.8% there was benzene [1]. Recently, Kuster et al. [2] have reported the presence of estrogens, progesterons and pesticides in the surface waters.

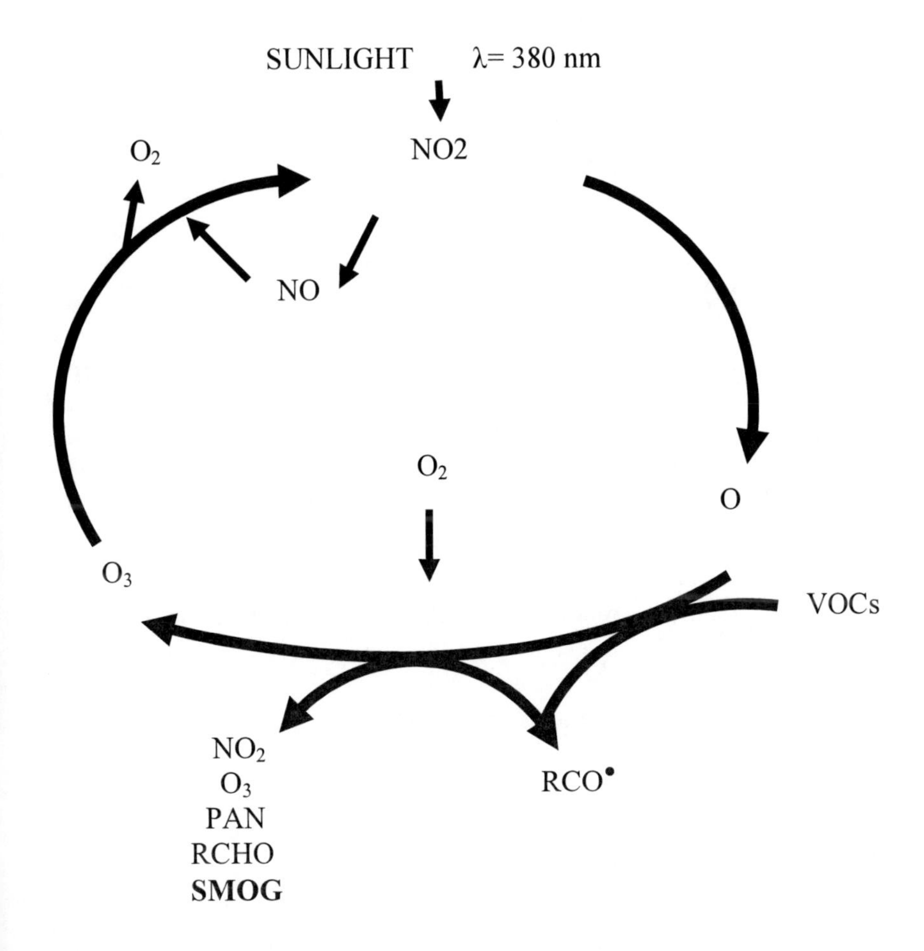

Figure 1. Esquema del proceso de formación del SMOG.

EFFECTS ON ENVIRONMENT AND HEALTH

The Photochemical Smog

In the summer of 1943, the city of Los Angeles underwent a phenomenon that was described as a suffocating fog provoking nausea, eye itching and respiratory airway problems. The subsequent scientific studies around 1950 demonstrated that the problem had resulted from a reaction between nitrogen oxides and VOCs in the presence of air and solar radiation giving rise to what is known as photochemical smog (*smoke* + *fog*). The reaction-forming smog is complex provided that, for instance, the fuel vapors have around 800 hydrocarbons that can participate in the process of smog formation. According to the Organization for Economic Cooperation and Development (OECD) records, vehicular combustion causes 44% of city VOCs such as benzene, 1,3-butadiene, H_2CO, benze(a)pirene, dibromide and ethylene dichloride ($(CH_2Br)_2$, $(CH_2Cl)_2$). The process may result from VOCs oxidation by a radical $HO^\bullet$ or by atomic O as observed in Figure 1 [3]. As expected, the interaction of NO_x, VOCs, sunlight and the formation of photochemical pollutants become important in big cities because of vehicular circulation, industrial activity and sunlight periods. USEPA recorded in 2005, as seen in Table 1, that the use of solvents, transports in general, process industry and other activities such as agriculture, gas stations, gas storage, and buildings are the main sources for VOCs generation in the USA.

The capacity of VOCs to form O_3 is calculated in relation to ethylene by means of the so-called POCP factor, Photochemical Ozone Creation Potential. This factor allows measuring the capacity to produce ozone from VOCs in function of ethylene kilograms. In Table 2, POCP values of some VOCs can be seen.

Table 1. Emissions of VOCs in USA*

Source	Total Emissions (Tons)
Vehicles	8358044
Use of Solvents	4245897
Industry of Processes	1645584
Other activities	1202517

* http://www.epa.gov/air/emissions/VOCs.htm

Table 2. Factors of photochemical ozone production

VOCs	PCOP
Alkenes	*84*
Aromatics	76
Aldehydes	*44*
Alkanes	42
Ketones	*41*
Alcohols	20

Regarding the environment, VOCs play a direct or indirect role in different actions:

1. Formation of tropospheric O_3: a phenomenon that affects human beings, animals and vegetation, in general.
2. Emissions of Greenhouse Effect: Most VOCs are rapidly oxidized but are secondary agents on the greenhouse effect due to the fact that they participate in the formation of tropospheric O_3, which has a potential as gas of greenhouse effect, 2000 times higher than CO_2, and in the formation of $HO^{\bullet}$ influencing in the oxidation and spatial distribution of methane.
3. Toxic or cancerous compounds: Some of them, such as toluene, are narcotics of the Central Nervous System, the long exposure to these compounds being very dangerous. Among those carcinogenic, we find benzene and 1,3-butadiene, are potential agents of leukemia, formaldehyde is a potential cancer agent in high respiratory airways and polynuclear aromatic hydrocarbons are lung cancer agents.
4. Depletion of stratospheric O_3: many VOCs exceed the tropospheric oxidation processes and reach high atmospheric levels. If they contain Cl or Br heteroatoms, these can participate in photolysis reactions that cause the depletion of the O_3 layer.
5. Accumulation and Persistence: Some VOCs of high molecular weight are more resistant to oxidation, being able to be dragged by rains and to form larger particles. Some of them can be bio-accumulative in animals and affect human health [4-6].

Indoor Pollution

Human beings at working age stay indoors (homes, offices, workshops, etc.) about 75% of the day, breathing air which may reach pollution values 10 times higher than the air values outdoors; whereas, the most vulnerable population such as elderly, children and newly born babies spend more than 90% of the time indoors. Due to this fact, the indoor pollution analysis has recently become very important. The indoor pollutants can be classified into two categories, taking into account the damage they produce in human beings: those associated to acute illnesses (asthma, allergies, etc.) and those carcinogenic. Among the former we find microorganisms, CO_2, CO, NO_x, SO_x, VOCs and in the latter, heavy metals (Cd, Ni), benzene, N-nitrosamines, polycyclic aromatic hydrocarbons (PAHs) among others. VOCs sources indoors are: cigarette smoke and fireplaces (formaldehyde, benzene, PAHs), from paints and synthetic materials (benzene, toluene, ketones, CCl_4), food cooking (acrylamide, acetaldehyde, and ethanol) and from cosmetics use and cleaning products such as perfumes, pharmaceutical products, deodorants and glass cleaners (chlorinated hydrocarbons, butane, propane, ethanol, ethyl acetate) [7-10].

Nowadays, one of the biggest problems is modern building construction where energy saving and comfort prevail but buildings lack purification systems and/or adequate air renovation systems. This situation led the World Health Organization (WHO), in 1982, to recognize that in a building where more than 20% of the population suffered problems such as headaches, eye, nose and throat irritation, fatigue, nausea, sleepiness, non-productive cough, dry or irritated skin, asthma, difficulties for concentrating and sensitivity to smells could be comprised in an illness called Sick Building Syndrome. According to this organization, the biggest source of this illness is cigarette smoke since CO, CO_2, particulate material (PM), gases and organic compounds such as formaldehyde, nitrosamines, aromatic hydrocarbons, hydrogen cyanide, ketones and nitriles are formed in its combustion.

On the other hand, VOCs of high molecular weight can condense and give rise to secondary organic aerosols (SOAs) formed from less volatile products of VOCs oxidation. In 2002, WHO determined that the long exposure to these kind of compounds and O_3 is related to the increase of lung and heart problems. Recently, WHO has pointed out that in Europe, between 1 and 5 babies in one thousand die from problems related to VOCs and their secondary products [11, 12].

Table 3. LEL values (%) of some volatile organic compounds (VOCs)

VOCs	LEL %
n-Butane	*1.8*
Benzene	1.3
Xylenes	*1.1*
CH_3OH	6.7
Methyl Ethyl Ketone	*1.9*
Acetone	2.6
Acetaldehyde	*4.0*

METHODS OF REDUCTION AND TREATMENTS OF VOCS EMISSIONS

The techniques for eliminating or reducing VOCs emissions are divided into primary and secondary. The former comprise aspects such as process change, process optimization and raw material change. These three methods are related to the Principles of Green Chemistry proposed by Anastas et al. [13]. Secondary techniques are those in which VOCs are recovered or destroyed. Within the recovery techniques we find Adsorption, Absorption, Membranes and Condensation. Within the destruction ones, we find biological and thermal techniques. In thermal techniques, it should be taken into account the lower explosivity limit (LEL). LEL is defined as the minimum concentration of VOCs in air or oxygen necessary for the mixture to be explosive. Table 3 shows LEL values of some VOCs.

As follows, some examples related to primary techniques will be developed and some of the main features of secondary techniques of recovery, bio-filtration and thermal incineration will be briefly mentioned. Finally, a more extensive development of oxidation or catalytic incineration will be described.

Primary Techniques

Regarding change and optimization of processes, new manufacturing technologies of adipic acid (AA) and polyolefins (PO) can be mentioned.

Regarding the process for obtaining AA, this acid is traditionally produced by the mixture cyclohexanol and cyclohexanone "KA oil" ("Ketone-alcohol").

The KA oil is oxidized with nitric acid for processing the adipic acid, using an excess of HNO_3. This process produces AA and esters which must be hydrolyzed to AA and other sub-products. In the modified process, cyclohexane reacts with O_2 in a tank stirred at low temperatures in presence of a Co catalyst dissolved in acetic acid. On the one hand, NO_x emissions are avoided and therefore its effect in photochemical reactions with VOCs. On the other, the system allows recycling cyclohexane and acetic acid avoiding VOCs emissions and reducing emissions and residues [14].

Nowadays, polymers are the chemical compounds most often used, the most important being propylene. Its production began in the 1950s from the development of Ti catalysts known as Ziegler-Natta catalysts. The original process resulted in Ti impurities in the polymers and they had to be purified with organic solvents with the consequent environmental impact. The new process has been designed by Basell. Spherizone Technology applies the theory of catalytic cracking in a multi-zone circulating catalyst. The polymer grains are in permanent circulation between two regions, each one having a different fluid dynamics. In a region, the solid particles move upwards by means of a gas current in rapid fluidization conditions, and then they separate from the carrier gas and cross to the second region in a downward movement. Through this effect, both regions may have different gas compositions and give rise to different materials, producing more uniform and homogeneous polymers. The change of technology according to the parameters of the company results in a reduction of VOCs emissions in 3000 kTon, reducing 50% CO_2 emissions and saving 15% of raw materials [15]. As it can be observed, the redesigned process can be economically and ecologically compatible.

Secondary Techniques

Choosing a secondary technique depends on several factors:

(a) Nature and concentration of VOCs
(b) Composition and Characteristic of VOCs mixture
(c) Waste Flow and Temperature
(d) Compliance Level of environmental standards
(e) Recycling possibility
(f) Possibility of using VOCs as additional energy source
(g) Cost of technology investment

The secondary techniques are classified in recovery methods (adsorption, condensation, membranes) and destruction methods (biofiltration, thermal and catalytic oxidation).

Recovery Technologies

Adsorption

It is based on the interaction adsorbate-adsorbent. The interaction can be physical (physisorption) or chemical (chemisorption). Depending on the solid system, zeolites or activated carbon, the efficiency can reach values between 90 and 96%, respectively. These systems work with VOCs concentrations between 50 and 10000 ppm, at room temperature, with flows between 0.05 to 3 m^3s^{-1}. They can work on vertical or horizontal beds. The cost of activated carbon is lower than that of zeolites, but these are advantageous for working in higher wet levels and they are not inflammable. In both solids, the adsorbed VOCs can be concentrated and destroyed or recovered according to the profits. In this sense, zeolite has a great thermal resistance allowing solvent recovery with high boiling point. [16,17]

Condensation

The driving force of the phase change is oversaturation, which is obtained by pressurizing and/or cooling down the polluted gaseous current. It depends on the boiling point of VOCs to be separated (in general close to 40 ºC) and the partial pressure of the VOCs that are dragged. It has 95% of maximum efficiency for concentrations higher than 10000 ppm which decreases to 50% if VOCs concentration is lower than 1000 ppm. The working flows are between 0.05 and 10 m^3s^{-1}. In the case of chlorinated VOCs, condensation columns with N_2*(l)* have been developed but they have some drawbacks, e.g. possibilities of overcooling and VOCs crystallization in some cases. In the case of using H_2O as fluid, maintenance is necessary to avoid scale formation. Operative costs can be very high if eliminating water from the condensate is necessary.

Separation by Membrane

The driving force of the separation by membranes can be a difference in pressure, in concentration or in electric potential. Generally speaking, the membranes can be classified as organic, normally polymeric, and inorganic (silica, alumina, zeolites, etc.). It is a technique sensitive to flow changes and

VOCs concentrations in the gaseous current, reaching an efficiency of 99%. Membranes work with flows between 0.1 and 1 m^3s^{-1} and have a thickness ranging from 2 to hundreds of micrometers. In reference to other recovery techniques, these membranes can work at temperatures up to 200 ºC. The polluted gas goes through the organic membrane and is condensed. Many systems use vacuum for keeping low partial pressure of solvents and then a preferential permeation takes place. The vapor is permeated through the membrane, compressed, cooled and condensed. The resulting vapors can be vented outdoors if their emission is in agreement with the environmental standards or treated by some destructive technique like incineration [18-19].

Destruction Technologies

Bio-filtration

The efficiency is function of the pollutant reaching values up to 95%. This technique can be used in a wide interval of VOCs concentrations between 15 and 5000 ppm. The high efficiency at low concentrations enables the bio-filtration to be suitable for eliminating dangerous VOCs, such as styrene and benzene, when they are found in gaseous currents in amounts of few ppm. An advantage of this technique is that it works at low temperatures between room temperature and 40 ºC. The maximum working flows are 10 m^3s^{-1}. One of its drawbacks is that it is necessary to keep the nutrient balance (C, N, P) in gaseous currents. Another disadvantage is that there are no VOCs able to poison microorganisms, for instance, by HCl formation. Besides, biodegradation can be different for equal microorganisms in a current of different VOCs.

In some cases, the products resulting from bio-filtration cannot oxidize VOCs completely. To set an example, the bio-filtration of ethylene trichloride can give rise to vinyl chloride as a sub-product whose toxicity is higher than the agent to be eliminated [20, 21].

Incineration or Oxidation

The incineration techniques comprise the thermal and catalytic incineration. Both techniques have applications in paint, printing and off-set factories and in industries where solvents are used. The *thermal incineration* can be classified as Recovering and Regenerative.

The former consists of a cylindrical block surrounded by a tubular reactor concentric to the block. The efficiency depends on the kind of oxidation temperature of VOCs in the mixture (in general, it is higher than 800 ºC), on the residence time, whose value is between 0.5 and 1 s and on the turbulence that allows an adequate distribution of temperature and O_2-VOCs mixture. In this kind of system, the flow can have maximum values of 10 m^3s^{-1}. VOCs concentrations can be up to 50 mgm^{-3} and the energetic efficiency can reach 70%.

The regenerative thermal incineration consists of a series of ceramic beds together with the combustion chamber and the burner. The number of beds depends on the volume to be treated and the system allows the current heating previous to oxidation and the subsequent cooling of combustion gases taking advantage of the heat exchange for the pollutant gases pre-heating. It works with larger flows than the recovering technique, being able to operate up to 100 m^3s^{-1}. Due to high transference, constant temperature and longer residence times, the energetic efficiency can reach 95%.

The catalytic incineration or oxidation is more advantageous than the thermal one since it works at temperatures lower than 450 ºC, though it depends on the catalyst. It can work at flows lower than 1 m^3s^{-1} and up to 10 m^3s^{-1}. Moreover, it can eliminate pollutants in low concentrations, thus this technique allows treating indoor pollutants. Kolaczkowski [22] points out that the combustion temperature depends on different factors such as (i) the chosen catalytic system, (ii) the catalyst load, (iii) the gas rate on the bed, (iv) working pressure and (v) bed length, (vi) concentration of each VOCs in the gaseous current. On the other hand, it should be taken into account that the emissions from fixed VOCs sources are usually transients which can cause trouble as regards to the efficiency for elimination, for instance.

The chemical structure of VOCs is one of the variables to bear in mind when choosing the catalyst. Schwartz et al [23], when studying energy of the oxidation activation of different organic molecules on Pt and Pd, have pointed out that the reactivity is related to the weakest bond energy of the molecule. Hernia and Vigneron [24] have stated from the oxidation study of different VOCs on Pt/Al_2O_3. They have reported the following order of VOCs destruction:

Alcohols> Esters > Aldehydes > Alkenes> Aromatics > Ketones > Esters > Alkanes

As observed, the destruction is related to the nature of the organic compound. O´Malley et al. [25] have determined that oxidation of oxygenated organic species such as alcohols and aldehydes has two mechanisms of combustion to CO_2 depending on the dissociation enthalpy of the weakest bond. By means of DRIFT spectroscopy, Peluso et al [26] correlate the catalytic activity with the adsorbed species in ethanol oxidation on MnOx, finding that the aldehyde is easily oxidized to CO_2, whereas the species alcohoxy are strongly adsorbed and need higher temperatures to be fully oxidized. In agreement with these results, Batiot [27] and Blausim-Aubè [28] show that the oxidation from acetaldehyde to CO_2 is energetically more favorable.

In saturated hydrocarbons, the reactivity is determined by the dissociation energy of the weakest CH bond. Besides, the breakage of a CH bond is easier for a secondary carbon than for a primary one [29]. In the case of alkenes, these can be adsorbed without dissociation on metal by means of bond π electrons, with an organometallic compound being formed that favors the breakage of the CH bond. Moro-Oka et al. [30] state that the greater activity of acetylene than that of ethylene and propane observed on noble metals is due to the presence of triple bonds.

The oxidation, combustion or catalytic incineration have their origin in the catalyst development for cars, that is why the first studied systems were noble metals such as Pt and Pd supported on Al_2O_3 and SiO_2. Nowadays, active phases like noble metals such as Pt, Pd, Ru and Au and oxides of transition metals such as Mn, Fe, Co, V, Cu on different supports such as Al_2O_3, SiO_2, CeO_2, TiO_2, zeolites (ZSM5, Y) are widely studied and used in different purification processes of gaseous currents.

Regarding catalysts based on noble metals, Garetto et al [31] have studied benzene combustion on Pt/Al_2O_3 analyzing the amount of Pt on the oxidation. From T_{50} measurement (temperature at which 50% of VOCs conversion occurs), the authors determine that larger Pt particles make benzene oxidation easier, due to the fact that density increases in oxygen superficial sites. Similar results have been reported for cyclopentane and CH_4 oxidation on the same catalytic system [32].

Scirè et al [33] have studied oxidation of methanol, 2-propanol and toluene on metals of group 11(IB) supported on Fe_2O_3 in excess of oxygen. The authors determine an Au>> Ag >Cu reactivity order, which is associated to the fact that the metal easily weakens the bond Fe-O and favors oxygen mobility in the oxidation process. The oxidation of 1-hexane, benzene and 2-propanol has been studied by Centeno et al [34] on $Au/CeO_2/Al_2O_3$ and Au/Al_2O_3 catalysts. These authors observe a better catalytic activity in the solid with CeO_2, and this may be due to an increase in the mobility of labile oxygens. In addition, they point out that the presence of high Au dispersion favors the formation of active sites for alcohol oxidation and that the presence of CeO_2 avoids the formation of propene and enable the oxidation of intermediate species like acetone. Okal et al [35] have assessed Ru as catalyst for VOCs elimination. Using butane as a molecule test, the authors state that the activity is favored by the presence of thin metallic Ru layers together with non-stoichiometric RuO_2, and that the temperature increase leads to RuO_2 formation, deactivating the catalyst. Recently, Bazhinimaev et al [36] have developed a new catalytic support from glass fibers (GFC), impregnated with a Pt salt with a concentration of 0.02 wt%. The catalyst is compared to a traditional Pt/Al_2O_3 catalyst in ethyl benzene oxidation, finding that the Pt/GFC catalyst is more active due to the greater Pt dispersion on glass fibers.

Generally, the catalysts based on noble metals have relatively low total oxidation temperatures, which represents energy saving but the cost of active phases is high. That is why the study of transition metal oxides has gained importance due to the resistance of these oxides to poisons like Cl and to their low cost, even though they work at slightly higher temperature than noble metals.

Rivas et al [37] have developed mixed oxides of Ce and Zr in different proportions, studying their activity in toluene and Cl-VOCs elimination. The authors point out that ZrO_2 concentrations higher than 50% favor Cl-VOCs oxidation due to a synergic effect of acid sites and to the lability of lattice oxygens. Also, they demonstrate that the solids with low ZrO_2 content favor toluene oxidation. In reference to the analysis of VOCs mixture, the authors have determined that toluene has a positive effect in Cl-VOCs oxidation since it favors HCl formation instead of Cl_2 molecule that is formed when studying Cl-VOCs individually.

Different authors have pointed out that VOCs oxidation on transition metal oxides may result in sub-products that in some cases could become more toxic than the one to be eliminated. Lintz and Wittstock [38] have analyzed the total oxidation of i-propanol on different types of oxides. These authors refer to the formation of sub-products such as formaldehyde, acetaldehyde and acroleyne on oxides such as Co_3O_4, Cr_2O_3, CuO, MnO_2, V_2O_5, NiO and γ-Al_2O_3.. In the same sense, as mentioned above, Peluso [26] in the study on ethanol oxidation on MnO_x has pointed out the formation of acetaldehyde at low ethanol conversions.

Lately, the use of Mn catalysts has become of great scientific interest. These catalysts are used as oxides [39 - 41], perovskites [28, 42, 43] or Cu spinels [44 - 46].

Figure 2. (a) Metallic Monolith of Al; Al_2O_3/Al (right) and with different charges of Mn (central and left); (b) Al Foam.

In general, MnO_x are excellent catalysts in the oxidation of oxygenated VOCs such as alcohols, aldehydes, ketones and esters. Their capacity to oxide VOCs is attributed to (i) the existence of the redox pair Mn^{3+}/Mn^{+4}, (ii) the poor crystallinity of oxides, (iii) Mn^{+4} vacancies [26, 41, 47].

The catalytic activity of perovskites in oxidation reactions of VOCs depends either on the molecule or on the oxidation mechanism involved. Voorhoeve et al. [48] have proposed two stages in the oxidation mechanism of VOCs. In the first stage, the molecule adsorption occurs on oxygen vacancies and in the second, the metallic cation intervenes in the redox process of oxidation. Different authors [49, 50] have shown that perovskites of $La_xMn_yO_z$ doped with Pd and Ag have a high catalytic activity in the oxidation of toluene, n-heptane, ethanol and Cl-VOCs. On the other hand, the hopcalite, Cu spinel ($CuMn_2O_4$), is a very active solid in the reduction of VOCs in an amorphous state. Its catalytic performance becomes poor when the temperature increases, since the hopcalite becomes more crystalline. Vu et al [46] point out that this kind of spinel is very active in VOCs oxidation when the stoichiometry is not achieved, enabling oxygen mobility in the lattice and therefore, the combustion.

Recently, Debecker et al. [51] have developed an $Ag\text{-}V_2O_5/TiO_2$ bifunctional catalyst, which has been assessed in the benzene total oxidation. The method of Ag impregnation on V_2O_5/TiO_2 leads V to promote autocatalytically the surfactant oxidation allowing the formation of Ag nanoparticles that are not syntherized. This Ag-V synergy makes the benzene conversion reach a value of 70% at 350 °C.; whereas in the V_2O_5/TiO_2 solid, the conversion reaches a value of 30% at the same temperature.

At pilot and industrial plant level, the catalytic elimination systems of gaseous currents work on fixed beds. One of the worst problems is that the use of pellets results in the pressure drop where the particle diameter (D_p), according to Ergun equation, is the most influencing factor since the pressure decrease is inversely proportional to D_p. For solving this problem, structured systems have been developed such as foams and monoliths, being able to be both metallic (Figure 2) or ceramic systems [52 – 54].

The preparation techniques of metallic foams can be from molten metal, by electrodeposition and vapor deposition; whereas the ceramic foams can be formed by replication, polymerization, chemical deposition and expansion. The metallic monoliths depending on the metal (Al, Steel, Cu) can be manufactured by electrolysis or thermal treatments. It is important to take into account that for manufacturing the catalytic system, it should be considered: (i) metallic and thermal properties of the structure, (ii) generating a surface on the structure able to anchor the active phase (catalyst), (iii) the collocation technique of the active phase on the structured support in order to avoid generating another kind of phase, which may not be active, (iv) the active phase-support adhesion, (v) the coating homogeneity [55, 56]. The coating of the structured support can be carried out by different techniques: washcoating, superficial growing and chemical or physical deposit by vapor (CVD and PVD) [57, 58].

In short, the catalytic systems suitable for VOCs elimination can be noble metals, simple or mixed oxides of transition metals. The choice of the active phase will depend on the VOCs nature in the gaseous current. The development of catalysts from transition metals (Mn, Cu, Ce) should take into account how to give rise to solids with the redox pairs $M^{n+}/M^{(n+1)+}$, cationic vacancies and a poor crystallinity.

References

[1] "Fate of priority pollutants in publicly-owned treatment works" USEPA Report 440/1-82-003, 1982.

[2] Kuster M., Diaz-Cruz S., Rosell M., Lopez M., Barceló D., *Chemosp.* 2010, 79, 880-886.

[3] Leighton P. *"Photochemistry of Air Pollution"* Academic Press, New York, 1961.

[4] Guo H., Wang T., Simpson, I., Blake D., Yu X., Kwok Y., Li Y. *China Atm. Environm.* 2004, 38, 4551-4560.

[5] WHO *"Air Quality Guidelines for Europe"*, 2nd Edition, European Series, No. 91. 2000, Copenhagen.
[6] Lee S., Li W., Ao, C. *Atm. Environm.* 2002, 36, 225-237.
[7] Srivastava A., Joseph A., Devotta S., *Atm. Environm.* 2006, 40, 892-903.
[8] Wallace L., Pellizari E., Hartwell T., Davis V., Michael L. Whitmore R., *Environm. Research* 1989, 50, 37-55
[9] Tham K., Zuraimi M., Sckhar S., *Envrionm. Int.* 2004, 30, 1075-1088.
[10] *"Guidelines for indoor air quality. Selected pollutants"* WHO Regional Office for Europe, Copenhagen, 2010.
[11] The world health report 2020- Reducing Risk, Promoting Healthy Life http://www.who.int/whr/2002/en/whr02_ch4.pdf
[12] WHO Fifth Ministerial Conference on Environment and Health, 2010. http://www.euro.who.int/en/home/conferences/fifth-ministerial-conference-on-environment-and-health
[13] Anastas P., Warner J., *"Green Chemistry: Theory and Practice"*, Oxford University Press, New York, 1998.
[14] Decoster D., Vassiliou E., Dassel M., Rostami A. *"Methods of removing acetic acid from cyclohexane in the production of adipic acid"*, United States Patent 5929277, 1999.
[15] http://utenti.unife.it/luca.bani/ETHIC/Presentazioni/Covezzi.pdf
[16] Yu F., Luo L. Grevillot G. *J. Chem. Eng. Data* 2002, 47, 467-473.
[17] Boulinguiez B., LeCloirec P., *Energy Fuels* 2010, 24, 4756-4765.
[18] Kammarunddin H., Koros W., *J. Membrane Sci.* 1997, 135, 147-159.
[19] Bernardo P., Drioli E., Golemme G. *Ind. Eng. Chem. Res.* 2009, 48, 4638-4663.
[20] Kwon S-H., Cho D. *J. Ind. Eng. Chem.* 2009, 15, 129-135.
[21] Saravanan V., Rajamohan N., *J. Hazardous Mater.* 2009, 162, 981-988.
[22] Kolaczkowski S. "Treatment of volatile organic carbon (VOC) emission from stationary sources: catalytic oxidation of the gaseous phase" in *"Structured catalysts and reactors"* Cybulski A., Moulijn J. (Editors) Taylor Francis CRC, NY 2005, 147-170.
[23] Schwarz A., Holbrook L., Wise H. *J. Catal.* 1971, 21, 199-207.
[24] Hernia J., Vigneron S. *Catal. Today* 1993, 17, 349-356.
[25] O´Mailley A., Hodnett B. *Catal. Today* 1999, 64, 31-38.
[26] Peluso M., Pronsato E., Sambeth J., Thomas H., Busca G. *Applied Catal.* B 2008, 78, 73-79.
[27] Batiot C., Hodnett B., *Applied Catal.* A 1996, 137, 179-191.

[28] Blausim-Aubé V., Belkouch J., Monceaux A., *Applied Catal.* B 2003, 43, 175-186.
[29] Sokolovski V., *Catal. Rev. Sci. Eng.* 1990, 32, 1-49.
[30] Moro-Oka Y., Morikawa Y., Ozaki A., *J. Catal.* 1967, 7, 23-32.
[31] Garetto T., Apesteguía C., *Applied Catal* B 2001, 32, 89-96.
[32] Garetto T., Apesteguía C., *Catal. Today* 2000, 2-3, 191-198.
[33] Scirè S., Minico S., Crisafulli C., Galvagno S. *Catal. Commun.* 2001, 2, 229-232.
[34] Centeno M., Paulis M., Montes M., Odriozo*la J. Applied Catal.* A 2002, 234, 65-78.
[35] Okal J., Zawadski M., *Applied Catal.* B 2009, 89, 22-32.
[36] Balzhinimaev B., Paukshtis E., Vanag S., Suknev A., Zagoruiko A., *Catal. Today* 151, 2010, 195-199.
[37] De Rivas B., Gutiérrez-Ortiz S., López-Fonseca R., González-Velazco J. *Applied Catal. A* 2006, 314, 54-63.
[38] Lintz H., Wittstock K., *Applied Catal. A* 2001, 216, 217-225.
[39] Lamaita L., Peluso M., Sambeth J. Thomas H., Minelli G., Porta P. *Catal. Today* 2005, 107-108, 133-138.
[40] Peluso M., Gambaro L., Pronsato E., Gazzoli D., Thomas H., Sambeth *J. Catal. Today* 2008, 133-135, 487-492.
[41] Agüero F., Barbero B., Gambaro L., Cadus L. *Applied Catal.* B 2009, 91, 108-112.
[42] Sinquin G., Petit C., Libs S., Hindermann J., Kiennemann A. *Applied Catal.* B 2000, 27, 105-112.
[43] Schneider R., Kiessling D., Wendt G., Buckhardt W., Winterstein G., *Catal. Today* 1999, 47, 429-435.
[44] Morales M., Barbero B., Cadus L. *Fuel* 2008, 87, 1177-1183.
[45] Xingyi W., Qian K., Das L. *Applied Catal.* B 2009, 86, 166-175.
[46] Vu V., Belkouch J., Ould-Dris A., Tacuk B. *J. Hazardous Mater.* 2009, 169, 758-765.
[47] Santos V., Pereira M., Orfar J., Figuereido J. *Applied Catal. B* 2010, 99, 353-363.
[48] Vooorhoeve R., Remieka J., Trimble L. *Ann. N.Y. Acad. Sci.* 1976, 2, 272-276.
[49] Musialik-Piotrowsks A., Landmesser M., *Catal. Today* 2008, 137, 357-361.
[50] Musialik-Piotrowsks A., Syczewska K. *Catal. Today* 2002, 73, 333-342.
[51] Debecker D., Delaigle R., Joseph M., Favre Ch., Gaigneaux E., in 10th International Symposium "*Scientific Bases for the Preparation of*

Heterogeneous Catalysts" E. Gaigneaux, M. Devillers., Hermans S., Jacobs P., Marters J., Ruiz P. (Editors) Elsevier, 2010, Amsterdam.

[52] Wang L., Tran T., Vo D., Sakurai M., Kameyama H., *Applied Catal.* A 2008, 350, 150-156.

[53] Burgos N., Paulis M., Gil A., Gandía L., Montes M. "*Studies in surface science in catalysis"* 2000, 130 593-598.

[54] Sanz O., Echave J., Sanchez M., Monzón A., Montes M., *Applied Catal.* A 2008, 340, 125-132.

[55] Avila P., Garetto T., Montes M. "Estructura de catalizadores y adorbentes" Chapter 5, in *"Eliminación de emisiones atmosféricas de COVs por catálisis y adsorción"* T. Garetto, I. Legorburu., M. Montes (Editors) CYTED, 2008, Santa Fe.

[56] Twigg M., Webster D., "*Metal and Coated Metal Catalysts" in "Structured catalysts and reactions"* Cybulski A., Moulijn J. (Editors) Taylor Francis CRC, NY 2005, 71-108.

[57] Xu X., Moulijn J. "Transformation of a structure carrier into a structured catalyst" in *"Structured catalysts and reactions"* Cybulski A., Moulijn J. (Editors) Taylor Francis CRC, NY 2005, 751-778.

[58] Avila P., Montes M. Miro E. *Chem. Eng. J.* 2005, 109, 11-36.

In: Volatile Organic Compounds
Editors: J. C. Hanks et al. pp. 167-175
ISBN 978-1-61324-156-1

Chapter 6

IN-VIVO ANALYSIS OF PALM WINE (ELAEIS GUINEENSIS) VOLATILE ORGANIC COMPOUNDS (VOCS) BY PROTON TRANSFER REACTION-MASS SPECTROMETRY

Ola Lasekan[2,*] and Sabine Otto[1]
[1]Deutsche Forschungstanstalt fuer Lebensmittelchemie, Lichtenbergstrasse 4, D-85748 Garching, Germany
[2]Universiti Putra Malaysia, 43400, Serdang,Selangor, Malaysia

ABSTRACT

The in-vivo volatile organic compounds (VOCs) release patterns in palm wine was carried out using the PTR-MS. In order to analyze the complex mixtures of VOCs in palm wine, the fragmentation patterns of 14 known aroma compounds of palm wine were also investigated. Results revealed masses m/z (43, 47, 61,65,75,89 and 93) as the predominant ones measured in-breathe exhaled from the nose, during consumption of palm wine. Further studies of aroma's fragmentation patterns, showed that the m/z 43 is characteristic of fragment of various compounds, while m/z 47 is ethanol, m/z 61(acetic acid), m/z 65 (protonated ethanol cluster ions), m/z 75 (methyl acetate), m/z 89 (acetoin) and m/z 93 (2-phenylethanol) respectively. The dynamic release

* Corresponding author (lasekan61156@yahoo.com), Fax: 03-8942 3552.

parameters (I_{max} and t_{max}) of the 7 masses revealed significant (P=0.05) differences, between maximum intensity (I_{max}) and no significant (P=0.05) differences between t_{max} among VOCs respectively.

Keywords: PTR-MS, Palm wine, Volatile Organic Compounds, Fragmentation Patterns.

Industrial relevance: This study is of relevance because PTR-MS is a very fast and powerful tool for obtaining real – time data for example in process control and time intensity studies. Also, the PTR-MS has the capability to detect low odor threshold compounds that are perceived by the human nose but are not detected by FID or by any other instrumental detector.

INTRODUCTION

Palm wine also called palm Toddy or simply Toddy is an alcoholic beverage created from the sap of various species of palm tree. Palm wine is a refreshing beverage enjoyed by people in parts of Africa, Asia and South America [1]. It has a milky flocculent appearance due to its high concentration of yeast which thus serves as a rich dietary source of vitamins of the 'B' complex [2]. For instance an estimated 225 x 10^6 litres is believed to be consumed in Nigeria annually [3]. From a consumer perspective, the most appealing features of palm wine are its flavor, and nutrition. Food flavor appreciation is one of the first evaluation signals along with food appearance and texture encountered by consumers during eating of food [4]. Furthermore, it is well known that this food's characteristic strongly influences consumer acceptability/preference judgment. Information on the flavor chemistry of natural and synthetic palm wine during production and consumption is increasingly important for the palm wine industries in determining optimal production and maintenance of acceptable flavors.

Also, an objective of palm wine research is to find chemical markers for the wine, which correlate with the sensory assessment of expert wine tasters and which drive consumer preferences. Currently, controls on aroma are still being carried out by human experts, who check and judge the bouquet of wine by perceiving the volatile compounds in the headspace. The need for an automatic technique which can reproduce the sensitivity of the human nose is highly important. One very interesting technique with rather high sensitivity and/or which can be used for this purpose is the proton transfer reaction- mass spectrometry (PTR-MS). PTR-MS is a relatively new MS technique which basically implements H_3O^+ as the ionizing agent in a chemical ionization [5]. The advantage of using H_3O^+ as the primary ion lies in its non – dissociative proton transfer reaction with most volatile organic compounds, whereas it does not react with any natural components of air. Compared to conventional MS, the non – dissociative character of the proton transfer reactions leads to less complex spectra and opens the opportunity to skip a previous separation of the compounds [5]. Detection limits of a few ppbv allow an analysis of the headspace without any previous concentration step. Therefore, PTR-MS is a very fast and powerful tool for obtaining real – time data for example, in process control and time intensity studies [6]. It has been used for food analysis, especially for the determination of flavor volatiles [6].Recently, PTR-MS has been used to evaluate volatiles in the head space of extra virgin and rancid olive oils in order to detect oxidative alterations [7] and geographical origin classification [8] of olive oils respectively. PTR-MS has variously been used in the classification of butter and butter oil [9], identification of strawberry cultivars [10] and to predict the sensory profile of espresso coffee [11]. The role of PTR-MS as a useful on-line monitoring tool for benzene formation in stored beverages [12] has also been reported. Furthermore, the capability of the PRT-MS method to detect low odor threshold compounds that are perceived by the human nose but are not detected by FID or by any other instrumental detector has been reported [13] .

In the present study, PTR-MS technique is applied to define the flavor profile of in – breathe exhaled from the nose during consumption of palm wine. To analyze the complex mixtures of volatile compounds in palm wine, basic data on the behavior of individual compounds is essential. Therefore, the fragmentation patterns of 14 individual volatile flavor compounds earlier identified in palm wine [14] were first evaluated.

2. Materials and Methods

2.1. Materials

Three bottled palm wine (4% vol, 1.5 L) were purchased at a wine shop in Nigeria and dispensed into a hundred 45 mL glass – tubes and stored at -20°C until analysis.

2.2. Panelists

Four panelists (two males, two females, ages 22 – 40 years, non smokers) were from the Technical University of Munich. They exhibited no known illnesses at the time of examination and normal olfactory and gustatory function. In regular weekly training sessions, panelists were tested for their sensory performance with selected suprathreshold aroma solutions prior to participation in the experiments, whereas subjective aroma perception was tested with a defined set of aroma substances.

2.3. Proton Transfer Reaction- Mass Spectrometry of Palm Wine

PTR-MS (Ionicon Analytik, Austria) has been described in detail elsewhere [6], [12], [15]; therefore, only a brief summary will be given here. The key elements of successful operation of PTR-MS are: an intense source of H_3O^+ primary ions, giving 10^6 counts S^{-1} primary ions; a cross – section for proton transfer which ensures unit conversion efficiency of primary ions to secondary ions on every collision; a mass analyzer which selects the ion peaks specific to the trace compounds present in the exhaust gas; and an electron multiplier (EM) detector operated in pulse counting mode with single particle detection efficiency. By optimizing the characteristics of all these parts, we achieved real time sub – ppb detection sensitivity of VOCs in air. As confirmed by the experiments, the formation of spurious chemical compounds due to reactions occurring inside the drift tube could be considered negligible unless the concentration of the compounds under study becomes large. The quantity measured with PTR-MS is usually the intensity of a protonated compound, on the mass of which information is obtained. This does not directly allow the definite identification of the compound itself because these are large numbers of compounds having the same nominal mass. Whenever qualitatively unknown mixtures of compounds have to be investigated, the problem of identification becomes a crucial one. Thus, PTR-MS is primarily a method for on – line monitoring of compounds rather than for gas analysis.

The headspace of a 20 mL palm wine aliquot contained in a temperature stabilized (29°C) 40 mL vial (Supelco) capped with a Teflon Septum was exchanged with ambient air with a continuous flow of 2 mL min^{-1}. This was flushed with nitrogen gas, and transferred through a heated capillary line directly into the reaction chamber. Five repeat measurements were performed.

The PTR-MS operated at standard conditions (drift tube voltage: 600 V, pressure: 730 mbar. U_2: 150 V, U_3: 80 V, SEM: 3050 V, inlet flows: 50 mL min^{-1}).

2.4. PTR-MS of Pure Volatile Organic Compounds (VOCs)

To assign character to the obtained ions from palm wine headspace, PTR-MS spectra of some pure VOCs were first determined. An aqueous solution (2 g L^{-1}) of the individual volatile compounds (10 mL) was placed in a glass vial (100 mL). The headspace was drawn at 50 mL min^{-1}, 15 mL min^{-1} of which was led into the PTR-MS. Samples were analyzed according to the method described by Lindinger et al. [6], and while employing a constant drift voltage of 600 V. Transmission of the ions through the quadrupole was considered according to the specification of the instrument. Background and transmission corrected spectra were averaged over five cycles. Presented PTR-MS spectra (Table 1) were obtained by normalizing the most abundant mass fragment to an intensity of 100.

2.5. *In Vivo* Flavor Release Analysis

Nose – space analysis aims at sampling and analyzing the air exhaled through the nose while palm wine is being consumed. For this, nose – space air was sampled via the two inlets of a glass nosepiece placed in both nostrils of the assessors. The nosepiece had one outlet for breathing and an orthogonal outlet for sampling. The latter was used to remove the air, without disturbing the assessor's breathing or swallowing pattern.

The air was drawn at a rate of 50 mL min^{-1}, 15 mL min^{-1} of which was drawn into the PTR-MS. The assessor placed his/her nose in the nosepiece with normal breathing for the first 30 s, after which the assessor drank 15 mL of palm wine and retained it in the mouth for another 30 s. Finally, the assessor swallowed the palm wine and continued to breathe into the nosepiece for another 60 s.

This was repeated three times for each of the four assessors (3x4 =12 replicates). The release of the predominant masses in the expired air (m/z 43, 47, 61, 65, 75, 89 and 93) was measured as described by Lindinger, et al. [6].

2.6. PTR-MS Data Analysis

Parameters calculated were; the maximum intensity of the released profile (I_{max}) and the time necessary to reach the maximum intensity (t_{max}). Using only the raw data obtained during consumption of palm wine until the swallowed event itself (Fig 1), the maximum intensity of this time interval represents I_{max}, and the time necessary to reach I_{max} is termed t_{max}.

3. Results and Discussion

3.1. Fragmentation Patterns in Palm Wine

To analyze the complex mixtures of volatile compounds in palm wine, basic data on the behavior of individual volatile flavor compounds earlier identified in palm wine [14] were first evaluated (Table 1). Generally, a high degree of fragmentation was observed. The protonated molecular ions showed the most abundant intensity for butanoic acid, 2/3 – methybutanoic acids, pentanoic acid, the esters (ethylbutanoate and methyl acetate), acetoin and ethanol respectively. 2- Phenyl ethanol split off water and break up to non – specific alkane fragment at m/z =93. Fragmentation of the aldehydes (3-methylbutanal, 3- methythiol propanal) resulted to m/z =39 and 49 as the dominant signals in the PTR-MS spectra of these compounds. On the other hand, the protonated molecular ion showed abundance intensity for two esters, i.e. ethylbutanoate, m/z = 117 and methyl acetate, m/z = 75 respectively. While the other ester, ethyl – 2- methylacetate had m/z = 103 as the dominant signal. Fragmentation of the ketones (2,3 – butandione and gamma – dodecalactone) led to m/z = 50 and 60 as the dominant signals in the PTR-MS spectra of these compounds.

3.2. PTR-MS Headspace of Palm Wine

A scan of the volatile compounds from the headspace of palm wine by PTR-MS revealed that the masses (m/z 43, 47, 61, 65, 75, 89, and 93) were the predominant signals obtained. Although, PTR-MS is a one – dimensional technique, the soft ionization results in either mass + 1 ion or characteristic large product ions. Studies on the breakdown patterns of spectra from the palm wine headspace showed that m/z 43 is characteristic for fragment of various compounds. While mass (m/z 47) is probably from ethanol. Mass (m/z 61) has *been reported to originate from acetic acid, 1 – propanol or fragment of ethyl* acetate [16]. On the other hand mass (m/z 75) is probably derived from methyl acetate, while mass (m/z 89) has been reported to originate from 2 – methyl propionic acid, butyric acid or acetoin [13].The last mass (m/z 93), is from 2 – phenyl ethanol.

3.3. The In – Mouth Situation

The intensities of those compounds in the breathe of the test persons were observed to change significantly after the consumption of palm wine are shown as dependent on time in figures 1 and 2b respectively. There are seven distinctly different groups of compounds as revealed by their masses. The most prominent of the compounds is ethanol (mass 47), followed by acetic acid (mass 61), fragment of many compounds (mass 43), protonated ethanol cluster ions (mass 65), acetoin (mass 89), methyl acetate (mass 75) and 2-phenyl ethanol (mass 93).These compounds rise to a maximum intensity shortly after ingestion of palm wine and decline to normal baseline values within the next 60 s (Fig 1). In the case of methyl acetate (mass 75), 2- phenyl ethanol (mass 93) and acetoin (mass 89) peak concentrations obtained from the breathe of the test persons were extremely low being < 1.0 x10 ppb respectively, while ethanol (mass 47) reached a peak concentration of approximately 14 x10 ppb in the breath of one of the test person (Fig. 2a). Recent investigations [17] showed that the amounts of odorants present in a food material or aroma solution can be considerably reduced in the food during consumption or mastication in the oral cavity. Thus, the observed low intensities of 2 – phenyl ethanol (mass 93) and methyl acetate (mass 75) in the breathe exhaled from the nose during consumption of palm wine could have been induced by adsorption or resorptive effects of the mouth mucosa or enzymic degradation of the odorants or, generally, interactions of odorants with salivary constituents. For instance, Buettner and Schieberle [18] reported that the adsorbed residue played an important role in the persistence of desirable or undesirable aroma in the mouth or in the nasal cavity. According to the present investigation, some of the adsorbed compounds were probably degraded by the enzymic activity of the saliva, thereby reducing their persistence as well as the overall intensity of the 'after smell' of these compounds, whereas, others might not be affected at all. It is worthy of note that the reason for the significant variation observed in the aroma breath profile of panelist 1 in respect of compounds with masses; 43, 47, 61 and 89 respectively, (Fig 2a), cannot be determined unambiguously. Finally, the time until maximum intensity is reached t_{max} is presented in figures 2b and 3b respectively. The t_{max} was higher for five compounds, with masses 75, 43, 47, 65 and 93, respectively. However, this effect was not significant (P=0.05)

The variations observed for the pulses between panelists (Fig 2a) indicated that the differences in the salivary activity of different humans exist, possibly resulting in a varying aroma perception. The dynamic release parameters (I_{max} and t_{max}) of the 7 masses (Figs 2 and 3) respectively revealed that while maximum intensity (I_{max}) differed significantly (P=0.05) among the compounds, there was no significant (P = 0.05) differences in t_{max} (which showed the persistence of a flavor compound).

CONCLUSION

The intensities of compounds in the breathe exhaled from test persons were observed to change significantly with time after the consumption of palm wine. Seven distinctly different groups of compounds with masses m/z (43, 47, 61, 65, 75, 89 and 93) were the predominant ones. The most prominent of the compound is ethanol (mass 47), followed by acetic acid (mass 61), fragment of many compounds (mass 43), protonated ethanol cluster ions (mass 65), acetoin (mass 89), methyl acetate (mass 75) and 2-phenyl ethanol (mass 93). The dynamic release parameters (I_{max} and t_{max}) of the seven masses revealed significant (P=0.05) differences, between maximum intensity (I_{max}) and no significant (P=0.05) differences between t_{max}. Finally, the coupling of PTR-MS with GC/MS would be necessary to allow for the quantization of the VOCs that contribute to a single PTR-MS ion signal.

REFERENCES

[1] L. Jirovetz, G. Buchbaucer, W. Fleischhacker, M.B. Ngassoum, *Ernahrung/Nutr.* 25(2001) 67.

[2] W.Van Pee, J.G.Swing, *East African Agric. Forestry J.* 36 (1971) 311.

[3] J.A.Ekundayo, J.E.Smith, D.R.Berry, B.Kristiansen (Eds). *An appraisal of Advances in biotechnology in Central Africa,* Academic Press, London, New York, 1980.
[4] F.Gasperi, G.Gallerani, A.Boschetti, F.Biasioli, A.Monetti, E.Boscani, A. Jordan, W.Lindinger, S.Iannotta, J.Sci. Food & Agric. 81 (2000) 357.
[5] A.Tani, S.Haywards, A.Hansel, C.N.Hewitt, *Int.J.Mass Spectrom.*239 (2004)161.
[6] W.Lindinger, A.Hansel, A.Jordan, *Int. J. Mass Spectrom. Ion Process,* 173 (1998) 191.
[7] E. Aprea, F. Biasioli, G. Sani, C. Cantini, T.D. Mark, F. Gasperi, *J. Agric. Food Chem.* 54 (2006) 7635.
[8] N. Araghipour, J. Colineau, A. Koot, W. Akkermans, J.M. Rojas, J. Beauchamp, A. Wisthaler, T.D. Mark, G. Downey, C. Guillou, L. Mannina, S. van Ruth, *Food Chem.* 1/108 (2008) 374.
[9] S. van Ruth, A. Koot, W. Akkermans, N. Araghipour, M. Rozijn, M. Baltussen, A. Wisthaler, T.D. Mark, R. Frankhuizen, *Eur. Food Res. Technol.* 1/227 (2008) 307.
[10] P.M. Granitto, F. Biasioli, E. Aprea, D. Mott, C. Furlanello, T.D. Mark, F. Gasperi, *Sensors and Actuators B: Chemical* 2/121 (2007) 379.
[11] C. Lindinger, D. Labbe, P. Pollien, A. Rytz, M.A. Juillerat, C. Yeretzian, I. Blank, *Anal. Chem.* 12 (2008) 215.
[12] E. Aprea, F. Biasioli, S. Carlin, T.D. Mark, F. Gasperi, *Int. J. Mass Spectrom,* 1-3/275 (2008) 117.
[13] E.Boscaini, T.Mikoving, A.Wisthaler, E.VonHartungen, T.D.Mark, Int. *J. Mass Spectrom,* 239 (2004) 215.
[14] O.Lasekan, A.Buettner, M.Christlbuer, *Food Chem.* 105 (2006) 15.
[15] E. Aprea, F. Biasioli, G. Sani, C. Cantini, T.D. Mark, F. Gasperi, Rivista *Italiana Sostanze Grasse* 2 (2008) 106.
[16] E.Boscaini, S.VanRuth, F.Biasioli, F.Gasperi, T.D.Mark, *J.Agric. Food Chem.* 51 (2003) 1782.
[17] A.Buettner, *Food Chem.* 50 (2002) 3283.
[18] A.Buettner, P.Schieberle, *J. Agric. Food Chem.* 49 (2001) 2387.

INDEX

A

B

C

D

E

F

G

H

I

J

K

L

M

N

O

P

T

U

V

W

X

Y

Z